Recent Progress
in the Study of Variation,
Heredity, and Evolution

ROBERT HEATH LOCK

CAMBRIDGE UNIVERSITY PRESS

Cambridge, New York, Melbourne, Madrid, Cape Town,
Singapore, São Paolo, Delhi, Mexico City

Published in the United States of America by Cambridge University Press, New York

www.cambridge.org
Information on this title: www.cambridge.org/9781108059626

© in this compilation Cambridge University Press 2013

This edition first published 1906
This digitally printed version 2013

ISBN 978-1-108-05962-6 Paperback

CAMBRIDGE LIBRARY COLLECTION

Books of enduring scholarly value

Life Sciences

Until the nineteenth century, the various subjects now known as the life
sciences were regarded either as arcane studies which had little impact
on ordinary daily life, or as a genteel hobby for the leisured classes. The
increasing academic rigour and systematisation brought to the study of
botany, zoology and other disciplines, and their adoption in university
curricula, are reflected in the books reissued in this series.

Recent Progress in the Study
of Variation, Heredity, and Evolution

In the nineteenth century and beyond, scientists at Cambridge produced
some of the most significant developments in the study of biological
variation and inheritance. The work of William Bateson (several of whose
books are also reissued in this series) was especially important in this
regard. This book, first published in 1906 by the botanist Robert Heath
Lock (1879–1915), lucidly traces these and other milestones in modern
biological understanding. A readable account is given of the evolution
of the discipline since the publication of Darwin's *On the Origins of Species*
in 1859, taking in the biometrical contributions of Francis Galton and
the research into mutation conducted by Hugo de Vries. The pioneering
experiments of Gregor Mendel, and the more recent rediscovery of his laws
of inheritance, are clearly contextualised so that non-specialist readers can
appreciate the scientific progress that had been made in the half-century
prior to the book's first publication.

Cambridge University Press has long been a pioneer in the reissuing of out-of-print titles from its own backlist, producing digital reprints of books that are still sought after by scholars and students but could not be reprinted economically using traditional technology. The Cambridge Library Collection extends this activity to a wider range of books which are still of importance to researchers and professionals, either for the source material they contain, or as landmarks in the history of their academic discipline.

Drawing from the world-renowned collections in the Cambridge University Library and other partner libraries, and guided by the advice of experts in each subject area, Cambridge University Press is using state-of-the-art scanning machines in its own Printing House to capture the content of each book selected for inclusion. The files are processed to give a consistently clear, crisp image, and the books finished to the high quality standard for which the Press is recognised around the world. The latest print-on-demand technology ensures that the books will remain available indefinitely, and that orders for single or multiple copies can quickly be supplied.

The Cambridge Library Collection brings back to life books of enduring scholarly value (including out-of-copyright works originally issued by other publishers) across a wide range of disciplines in the humanities and social sciences and in science and technology.

Selected titles by Darwin and other participants in the early debates on evolutionary theory, available in the
CAMBRIDGE LIBRARY COLLECTION

Candolle, Augustin Pyramus de, and Sprengel, Kurt: *Elements of the Philosophy of Plants* (1821) [ISBN 9781108037464]

Cuvier, Georges, translated by Robert Kerr: *Essay on the Theory of the Earth* (1815) [ISBN 9781108005555]

Darwin, Charles: *Geological Observations on South America* (1846) [ISBN 9781108027144]

Darwin, Charles: *Geological Observations on the Volcanic Islands, Visited During the Voyage of H.M.S. Beagle* (1844) [ISBN 9781108072335]

Darwin, Charles: *Insectivorous Plants* (1875) [ISBN 9781108004848]

Darwin, Charles: *Journal of Researches into the Natural History and Geology of the Countries Visited during the Voyage of H.M.S. Beagle* (second edition, 1845) [ISBN 9781108038065]

Darwin, Charles: *Monographs on the Fossil Lepadidae, Balanidae and Verrucidae* (1851) [ISBN 9781108004824]

Darwin, Charles: *On the Movements and Habits of Climbing Plants* (1865) [ISBN 9781108003599]

Darwin, Charles: *On the Various Contrivances by which British and Foreign Orchids are Fertilised by Insects* (1862) [ISBN 9781108027151]

Darwin, Charles: *The Different Forms of Flowers on Plants of the Same Species* (1877) [ISBN 9781108018272]

Darwin, Charles: *The Effects of Cross and Self Fertilisation in the Vegetable Kingdom* (1876) [ISBN 9781108005258]

Darwin, Charles, edited by Francis Darwin: *The Expression of the Emotions in Man and Animals* (second edition, 1890) [ISBN 9781108004831]

Darwin, Charles: *The Formation of Vegetable Mould through the Action of Worms* (1881) [ISBN 9781108005128]

Darwin, Charles, edited by Francis Darwin: *The Life and Letters of Charles Darwin* (3 vols., 1887) [ISBN 9781108003421]

Darwin, Charles: *The Origin of Species* (sixth edition, 1872) [ISBN 9781108005487]

Darwin, Charles: *The Structure and Distribution of Coral Reefs* (1842)
[ISBN 9781108065627]

Darwin, Charles: *The Variation of Animals and Plants under Domestication*
(2 vols., 1868) [ISBN 9781108014243]

Darwin, Charles: *The Descent of Man and Selection in Relation to Sex* (2 vols., 1871)
[ISBN 9781108005111]

Darwin, Charles, edited by Francis Darwin: *The Foundation of the Origin of Species*
(1909) [ISBN 9781108004886]

Darwin, Charles, edited by Francis Darwin: *The Power of Movement in Plants* (1880)
[ISBN 9781108003605]

Darwin, Charles, Henslow, John Stevens, and Sedgwick, Adam:
The Teaching of Science at Cambridge (1846) [ISBN 9781108002004]

Geikie, Archibald: *Charles Darwin as Geologist* (1909) [ISBN 9781108002578]

Lamarck, Jean-Baptiste Pierre Antoine de Monet de: *Philosophie zoologique*
(2 vols., 1809) [ISBN 9781108038041]

Lyell, Charles: *Principles of Geology* (3 vols., 1830–3) [ISBN 9781108001342]

Lyell, Charles: *The Geological Evidences of the Antiquity of Man* (1863)
[ISBN 9781108003971]

Romanes, George John: *Mental Evolution in Animals* (1883) [ISBN 9781108037877]

Romanes, George John: *Mental Evolution in Man* (1888) [ISBN 9781108037976]

Romanes, George John: *Darwin, and after Darwin* (3 vols., 1893–7)
[ISBN 9781108038126]

Romanes, George John, edited by E.D. Romanes: *The Life and Letters of George John
Romanes* (1896) [ISBN 9781108037891]

Wallace, Alfred Russel: *Contributions to the Theory of Natural Selection* (1870)
[ISBN 9781108001540]

Wallace, Alfred Russel: *Darwinism* (1889) [ISBN 9781108001328]

For a complete list of titles in the Cambridge Library Collection please visit:
http://www.cambridge.org/features/CambridgeLibraryCollection/books.htm

VARIATION, HEREDITY, AND
EVOLUTION

Ch. Darwin

RECENT PROGRESS IN THE STUDY OF VARIATION, HEREDITY, AND EVOLUTION

By ROBERT HEATH LOCK, M.A.

FELLOW OF GONVILLE AND CAIUS COLLEGE, CAMBRIDGE

LONDON
JOHN MURRAY, ALBEMARLE STREET, W.
1906

PREFACE

THE idea of writing this little book occurred to me whilst reading Mr. W. C. D. Whetham's volume on 'The Recent Development of Physical Science.' I found the story of the modern progress of physics so interesting as to encourage the belief that a similar account of the subjects with which I was myself more particularly familiar might prove of a like interest to other people. I did not, indeed, suppose for a moment that I could vie with Mr. Whetham in the power of literary expression which renders his book so eminently readable. I rather hoped that the peculiar interest and importance of the theme might outweigh the present author's deficiencies in this respect.

For the group of subjects of which I intended to give a brief account Mr. W. Bateson has recently proposed the term 'genetics,' an expression which sufficiently indicates their scope to the initiated. Since, however, the meaning of the word 'genetics' is not yet clearly understood by everybody, it seemed better to adopt in the present instance a somewhat more descriptive title.

The rediscovery of Mendel's law some seven years ago led to a complete change in our attitude towards the problems of variation, heredity, and evolution ; and the new method of study thus introduced has rendered possible a renewal of the attack upon these problems with increased vigour and with remarkable results. At the present time this activity may be said to have its centre in the school of genetic research founded at Cambridge by the independent energy of Mr. Bateson. So far-reaching are the results already arrived at by Mr. Bateson and others, both in their scientific interest and in their probable influence upon human affairs, that it seemed desirable to give an immediate account of these and of kindred lines of recent study, even though the rapid progress which is a characteristic manifestation of this department of science must render any such attempt a more or less transitory one.

Whilst I was still engaged upon my task, the first volume of Dr. Lotsy's admirable ' Vorlesungen über Descendenztheorien ' made its appearance. But for the fact that most of the following pages had then already been written, I might have hesitated to pursue my project, since a book not altogether unlike the present might be produced by the comparatively simple process of making a series of judicious extracts from Dr. Lotsy's work. The latter is, however, in the German language, and on a considerable scale, so that there seemed still to be room for an introduction to

the science of genetics of the more modest dimensions
which I had contemplated. I should wish, however,
particularly to recommend Dr. Lotsy's lectures to any
reader who wishes to go further into these matters.

I am indebted to several friends for assistance during
the course of my work. Mr. R. P. Gregory kindly
read through the proof of the chapter on cytology; and
I wish here to record my thanks to Mr. J. Stanley
Gardiner, to Mr. C. T. Regan, to Mr. W. S. Perrin, and
to Mr. R. C. Punnett for information on special points.
To the last-named I owe the photograph which appears
as Fig. 15. I am particularly grateful to Mr. R. H.
Biffen and to Mr. G. Udney Yule for access to work
which has not hitherto appeared in print.

Adequately to acknowledge Mr. Bateson's influence
upon these pages is a more difficult matter, and not
the less so because I have deliberately refrained as
far as possible from consulting him whilst the book
was in course of preparation, in order that it might
retain if possible some traces of individuality. It is
therefore clear that he is in no way responsible for its
deficiencies. But, apart from the fact that I am
conscious of having quoted his ideas at more points
than could possibly be acknowledged seriatim, I owe
to Mr. Bateson both my first introduction to the science
of genetics, and a continual fund of encouragement in
the prosecution of studies connected with it.

I have to thank Mr. Francis Darwin for kind permission to reproduce a portrait of his father; Professor

de Vries for the present of an excellent portrait ; and Mr. Francis Galton for the loan of a photograph well known from earlier reproductions. The portrait of Mendel is reproduced from the frontispiece to Mr. Bateson's 'Defence,' by the permission of the Syndics of the Cambridge University Press. Messrs. Macmillan and Co. have kindly allowed the reproduction of the diagram which occupies p. 80, and of the table and figure on pp. 82 and 83. The figures facing pp. 136 and 143 are from de Vries' 'Mutationstheorie,' published by Messrs. Veit.

The attempt has been made to render the following pages intelligible to the general reader, as well as to the more scientific public, to which they are primarily addressed. A short glossary has been added, which may be found useful by those who have no previous acquaintance with biological terms.

CAMBRIDGE,
October 23, 1906.

CONTENTS

CONTENTS

CHAPTER VIII

MENDELISM (*continued*)

CHAPTER IX

RECENT CYTOLOGY

CHAPTER X

CONCLUDING CHAPTER

LIST OF ILLUSTRATIONS

PORTRAITS

DIAGRAMS

RECENT PROGRESS
IN THE STUDY OF VARIATION,
HEREDITY, AND EVOLUTION

CHAPTER I

INTRODUCTION

THE present volume deals with variation and inheritance in plants and animals, especially in so far as those subjects bear upon the problem of the origin of species. By inheritance we mean those methods and processes by which the constitution and characteristics of an animal or plant are handed on to its offspring, this transmission of characters being, of course, associated with the fact that the offspring is developed by the processes of growth out of a small fragment detached from the parent organism. The term ' variation,' on the other hand, includes a number of different phenomena which will be described at greater length as the work proceeds ; but, broadly speaking, we may say that the study of variation is concerned with the circumstance that members of the same species are not all alike, and more particularly with the fact that

I

differences are to be found between different members of the same family. Some of these differences arise comparatively late in life, and may be the result of circumstances or of education. It is the first duty of the student of variation to distinguish as far as may be possible between differences of this kind on the one hand, and those differences on the other which depend upon the fact that the different detached fragments, as we have termed them, of the parent organism—its germ-cells, in fact—show greater or smaller differences among themselves.

The facts of variation have this very special importance, that the whole theory of organic evolution is based upon them. The fact that members of the same species are not all alike, depending upon the further fact that offspring may differ from their parents, makes it possible in the course of generations for progressive changes to take place, so that from the offspring of different members of the same species different new species may arise. But for this fact of variation it would have been quite impossible for Darwin to have overthrown the former crude belief in a special creation of each separate species, since there would have been no material for his great factor—natural selection —to work upon. It is with variation, then, and with the manner in which characters appear in the successive generations of living things, that we are here concerned.

Ever since the publication of Darwin's 'Origin of Species' in 1859, these subjects, and especially the theoretical aspects of them, have been received even by the general public with all the signs of a genuine enthusiasm ; and none, moreover, can be more fascinating to the professional naturalist. But since the time of Darwin the more popular accounts have dealt almost exclusively with theoretical considerations and with matters of opinion. Highly abstruse controversies have raged freely between Neo-Lamarckians and Neo-Darwinians, and these ·have found a place in the pages of works ostensibly intended for the instruction of all and sundry ; whilst only a bare residuum of actual matters of fact has seen the light of popular publication. If the truth must be told, the experimental method was given up for a long time by the majority of specialists themselves in favour of the controversial, and, indeed, this tendency has by no means yet died out from among the habits of some professed evolutionists. On the other hand, during the last fifteen to twenty years, a few scattered workers have diligently applied themselves to a study of the facts of variation and inheritance, with results which already more than justify the anticipation in which their work was begun—namely, that by such methods alone can any real progress in our knowledge of the processes of evolution be brought about.

The science of organic evolution is by no means the simple and isolated study it might be supposed to be from a perusal of some of the more popular accounts. Its footing rests immediately upon the widest foundations which zoology, botany, and physiology can afford ; and these in their turn are ultimately based upon the results of chemical and physical science. But some

of the most fundamental parts of physical science, as
I think we may fairly call the branches of electricity
and molecular physics, seem at present to be undergoing modifications which bid fair to bring about a
complete revolution in current ideas upon these subjects. It is highly probable that these results will
ultimately lead to a considerable modification in pre-
vailing notions about living things ; but the new
developments have yet to reach biology through the
channels of organic chemistry, physiology, cytology,
and the like, and at present we do not know what the
result of this influx is likely to be. These considerations need not, however, detain us, for the new knowledge of variation and inheritance, of which it is proposed to give some account, is largely concerned with
the grosser characters of organisms, so that ultraminute
structures may be left alone for the present until the
stream of physical knowledge stirs them into greater
prominence. So much is this the case with the study
of variation and inheritance by experimental methods
at the present day, that this science is treated by some
with a fine contempt, because its tools are those of the
breeder and gardener, and because the assistance of
the compound microscope may often be laid aside for
days together. Yet this applies only to one aspect of
the subject, and the microscopic study of the embryonic
rudiments of organisms, going hand in hand with the
experimental observation of adult structures, is rapidly
leading to a clearer understanding of the processes of
heredity.

The problem which those who are engaged in this

kind of work have set themselves for solution is that of the nature and method of origin of the existing differences between certain groups of organic beings —namely, species. Basing their studies on the doctrine that the present species have arisen through the modification of pre-existing species, they endeavour to observe how modifications of existing species do actually arise in Nature, as well as under domestication ; and they watch the hereditary transmission of the modified forms when like is bred with like, and when different types are crossed together. For the theory of uniformity, now universally accepted, teaches us that the organisms with which we are now familiar owe their present characteristics to the accumulation of a series of changes similar to those which are still in progress. It has, therefore, appeared likely to a few that a further understanding of the processes of evolution might best be obtained by a closer study, firstly, of variation, or the ways in which offspring differ from their parents ; and, secondly, of inheritance, or the ways in which the resemblances between parents and their offspring are perpetuated from one generation to another.

It may be well to point out at once that the further study of the method of origin of new species, admitting, as it does, that this process is not yet by any means fully understood, does not for this reason imply that the theory of organic evolution itself is open to criticism. The evidence that new species arise by the modification of pre-existing species is quite independent of the evidence that this process invariably occurs by the action of natural selection upon minute differences, in the manner which Darwin described, and which has been claimed by others as the sole means by which the origin of new forms takes place. The evidences of evolution are much more numerous and more weighty than the evidences of the survival of the fittest. The theory of evolution, as opposed to the creation hypothesis, is supported by innumerable facts of classification, of morphology, and of embryology, by the geographical distribution of animals and plants, and by their succession in the geological strata, as well as by direct observation of the actual occurrence of changes in the case of domestic productions as well as under Nature, and many of these facts have no direct bearing upon the theory of natural selection.

Before discussing the problem of the origin of species, it is necessary to arrive at some idea as to what the term ' species ' means. And this is not altogether an easy matter, since a precise definition has not been, and cannot be, agreed upon. The idea of species is, indeed, of great antiquity and very gradual growth. Primitive men doubtless recognised certain plants or animals as being like one another, and different from others, and they gradually came to distinguish such forms by giving a different name to each. The names first used must have applied as a rule to genera rather than to species. Thus, such common names of plants as rose, bramble, vetch, nettle, dock, crowfoot, are names of genera—groups of greater extent than species, and often more easily defined than the latter. Later on civilized men paid closer attention to tne different kinds of plants, and the old herbalists discovered and described a number of different sorts of roses, of buttercups, and of other plants, and distinguished each by a descriptive sentence.

As more and more species came to be described, this method of designation became very cumbersome, until Linnæus, about the middle of the eighteenth century, adopted the idea of a binomial nomenclature (originally suggested by Bachmann), in which every species of each known genus received a separate name of its own to distinguish it, so that the different kinds of buttercups were now known as *Ranunculus acris*, *R. bulbosus*, *R. sceleratus*, and so on.

Linnæus himself appears to have had a very definite idea of what constituted a species, and in accordance with the view then current, he defined a species as being a group of organisms which owed its origin to a separate act of creation. From the nature of the case this definition could be of little use in practice. Practically, then, species were defined as groups of animals or plants, the members of which resembled one another in definite morphological characteristics—that is to say, in constant features of form and structure. This definition has survived the downfall of the dogma of the constancy of species, and at the present day species as defined by Linnæus are found to be groups of much merit both for naturalness and for convenience—at any rate in the case of plants. The fact that intermediate forms and minor groups do sometimes and to some extent bridge over the gap which separates a pair of species thus defined seems to have caused disquiet in the mind of Linnæus himself, and he recommended his disciples to have no dealings with these inferior varieties, as being beneath the dignity of a botanist to notice. Of late years these minor species have excited much attention, and it is to a study of this kind of species in particular that the mutation theory of de Vries owes its origin, as will be told in a later chapter.

Such minor groups, occurring within the limits of a single Linnæan species, and subdividing it into smaller collections of individuals, were made the object of special study in the case of plants by the French botanist, Jordan ; and for this reason they are sometimes referred to as Jordan's species. Jordan, for example—though the example is indeed an extreme one—described more than two hundred different types, all of which would formerly have been included in the single Linnæan species, *Draba verna*. To take a more familiar instance. We find in the ' British Flora ' of Bentham and Hooker the primrose, the cowslip, and the true oxlip, all described as varieties of one and the same species ; yet these three kinds of plants are now almost universally recognised to be as good species as any in nature.* In a similar way, on closer investigation, it has been found necessary to split up a considerable number of Linnæan species, and to subdivide each into several species of smaller range.

* A contrary opinion is, however, expressed in the *Journal of Botany* for July, 1906.

It has already been pointed out that Linnæus himself distinctly deprecated this process of splitting. 'Varietates levissimas non curat botanicus,' said Linnæus. Jordan, however, applied the method of experiment to many of the species of his own definition, and having transplanted them from a variety of localities to the uniform soil of a garden, found that they preserved their distinctive characters and came perfectly true to seed.

It appears then that Jordan's species are just such true and constant groups as those of Linnæus. They are separated from one another by definite features of form and structure, only these differences are not so wide as those which separate Linnæan species. The latter are, indeed, to be looked upon as more or less artificial groups or aggregates of these physiological species, as Jordan's species have also been called. The problem of the origin of the smaller groups is clearly to be placed before that of the origin of the larger species.

It is true that in the case of certain groups of animals and plants there would appear to be no possibility of drawing hard and fast lines between the species, which thus seem to shade gradually one into the other. There is, however, a great difference between the admission that certain nearly related species are difficult or impossible to separate definitely, and the statement that there is no true distinction between them, and the latter statement is one which few are bold enough to make. The case stands thus. We know that great numbers of large groups (classes

and families) of animals and plants exist, in which
the most nearly related species are quite definitely
distinct from one another. In other classes systematists have so far found great difficulty in framing
definitions of specific groups. We shall see later on,
though at first sight it may appear almost paradoxical,
that it is quite possible for groups to be perfectly
distinct, although individual members of them may
have deviated so far, each from its proper type, as to
render impossible the task of deciding from their
appearance which group any of these individuals
belong to.

Let us next consider a particular example of a class
of animals in which the discrimination of species is
difficult or impossible. This is said to be the fact
with the majority of sessile animals—such animals
as resemble plants in their stationary habit, and in
no case are the problems of species separation more
difficult than in the class of the stony corals. Now,
attempts to determine the species of corals have so
far been made almost entirely from a study of what
may be called vegetative characters—usually from
details of the shape and structure of the stony skeleton
of the animals. How far these features may be affected
by external circumstances has not been determined,
but it must be noted that the so-called skeleton is
entirely external to the living organism. Now we
know that in the case of many of the higher plants
vegetative characters are extremely liable to become
modified owing to the action of the environment.
Differences of moisture, light, soil, climate, and alti-

tude, are all capable of changing the general appearance of a plant so as to render it scarcely recognisable. Fortunately, in the case of the higher plants, the floral organs, which are the ones chiefly made use of for purposes of specific discrimination, are very little liable to modification by external conditions; but in the corals a similarly stable set of organs does not appear to have been discovered. It seems, therefore, hardly fair to regard the example of the corals as affording an established exception to what we must look upon as the general rule—namely, that species are on the whole definite and discontinuous groups.

As a rule, then, the species riddle presents itself definitely as the problem of the existence of a series of discontinuous groups of creatures, sharply marked off the one from the other, and often, too, existing among surroundings which afford no corresponding discontinuity, though each is well enough fitted for the life which it has to lead.

The problem which we have to face has been enunciated by Bateson in the form of the two following propositions :

' 1. The forms of living things are various, and on the whole are discontinuous or specific.

' 2. The specific forms on the whole fit the places they have to live in.

' How,' he continues, ' have these discontinuous forms been brought into existence, and how is it they are thus adapted ? This is the question the naturalist is to answer. To answer it completely he must find

(1) the modes and (2) the causes by which these things have come to pass.'

The differences between existing species are open to study in more than one way. By way of limiting the discussion for the present, we shall consider the case of plant species only ; but the methods of study which are applicable to animal species are of quite similar kinds.

Four methods at least are available. Firstly, that of comparison ; secondly, the method of statistical examination ; thirdly, the method of cultural experiment ; and lastly, that of cross-breeding.

The method of comparison is the one to which the ordinary worker in descriptive botany is almost of necessity confined. In this way plants which closely resemble one another are grouped together as belonging to the same species, and separated from others, the appearance of which is different. By appearance is not meant simply the general habit of the plants ; all morphological features whatever may be used for purposes of comparison, and the most minute are often of the greatest importance. But the systematist who works only in this way knows nothing of the real relationships between the plants with which he is dealing.

When a sufficient number of specimens is available, the methods of statistics can be applied. These involve the making of a series of accurate measurements or countings of the parts upon which depend the supposed differences or resemblances of the plants under consideration. The resulting numbers are then ranged in order so that a precise view of the numerical characters of a large number of specimens can be readily obtained. By the use of such methods valuable information is often to be arrived at. But the same limitation affects them as in the preceding case.

So that the only way in which we can come to a definite decision as to whether a given putative species does or does not represent a definite and constant type is by resorting to our third method, that of sowing its seeds and actually rearing its progeny. And this is not so simple a matter as might appear at first sight, for a great many precautions have to be taken. Thus we must separately sow the seeds of many different individuals of the species which we are examining, so as not to base our conclusions upon a few experiments only. But in many cases, even when this has been done, we should only know one of the parents of our seedlings—that is to say, in cases where the pollen for fertilization may possibly have been conveyed by natural agencies from a different plant. In such a case we must either ensure self-pollination by isolating our plants, or we must artificially provide pollen from a separate known parent. If under these circumstances a particular group of plants preserves the characteristic differences which distinguish it from another group which has also been grown for a number of generations under the same conditions, we have at last reasonable grounds upon which to base the opinion that we are dealing with two distinct physiological species, even though the visible differences between them may seem very small to an untrained eye.

Lastly, evidence of a confirmatory nature may be obtained by observing the results of cross-fertilization between a pair of closely allied species. Much, too, may be made out from the failure of such experiments, since the refusal of two plants to breed together is generally regarded as clear evidence of their specific distinctness. But for this reason the method of cross-breeding is more particularly adapted for the examination of forms somewhat nearly related to one another —for example, different members of the same species.

As the result of the methods presently to be described, the fact has been established that two entirely distinct sorts of divergencies may appear among members of a single family. Variations, that is to say, may be of two quite different kinds. In the first place we have those slight differences which invariably distinguish all the members of every family —individual variations which affect every part and every character. Such differences are known as fluctuating, normal, or continuous variations. As an example we may cite the variations in size or stature shown by the various members of any purely-bred race. When a large number of individuals are compared in respect of a character of this kind, they are found to fall into a continuous series ranging from a certain extreme of shortness on the one hand to an extreme of tallness on the other. Individuals of a medium height, however, are usually more numerous than either of the extreme forms. Some further account of the study of continuous variations will be given in Chapter IV.

The second kind of variation is variously known as abnormal, definite, and discontinuous variation, and includes what are known as sports and mutations. Such variations, as is indicated by the terms applied to them, involve definite differences usually of considerable amplitude. A good example of a discontinuous variation would be afforded by the appearance of a child having six fingers in a family in which this abnormality was not previously known to occur. We shall pursue the discussion of discontinuous variation and of the methods of perpetuation of the types which thus arise in Chapters V. and VII.

A short account of the historical development of the theory of organic evolution is given in Chapter II. It is of particular interest to notice that the modern view of the mutationist is foreshadowed with remarkable precision in the passage from Aristotle's ' Physics ' which is quoted in Chapter V. Passing to more recent times, three distinct accounts of the method of origin of specific differences have been proposed almost within the last century, and each of these theories still finds a number of supporters.

1. The view of Lamarck, published first in 1801, and in an enlarged form in 1809, was briefly as follows : Noticing that the organs of men and other animals are increased and strengthened by use, and particularly by conscious use, Lamarck assumed that this effect

could be passed on by inheritance from parent to offspring, and so accumulated from generation to generation. In the case of animals Lamarck conceived the production of a new specific form to take place in the following way : Owing to some change of external conditions, the desire to perform some new kind of action was set up in the parent species, and by the hereditary effect of the striving occasioned by this desire a modification of the organs affected into forms better fitted to carry out the new function was gradually achieved.

Thus Lamarck supposed that snakes were evolved from a pre-existing type of animal which was of a much less attenuated shape, and which possessed two pairs of limbs like any other vertebrates. And he supposed this evolution to have taken place owing to the constant striving of these animals to pass through narrow crevices ; the effect of such striving being inherited, and so accumulated from one generation to another.

In the case of plants, in which conscious effort is precluded, a similar result was supposed to have been attained by an hereditary accumulation of the effects of the environment.

2. The explanation of Darwin, or at least the Neo-Darwinian form of it, as interpreted by Wallace, Weismann, and others, and as opposed to and excluding the view of Lamarck, was as follows : Two separate factors are primarily concerned : (1) the fact of fluctuating variation—the fact that no two members of the same family ever resemble one another exactly ; and (2) the occurrence of a struggle for existence between organisms owing to the geometric rate of

increase of living things. From these two facts it follows that when a change of environment takes place, certain members of an existing species will be somewhat better adapted than others to withstand the new conditions, and the former will tend to survive to the exclusion of the latter. It is assumed that during a long series of generations this process will cause a steady change in the character of the species in the direction of better adaptation to the new conditions.

Thus we might suppose that among the ancestors of the snakes those which happened to possess the longest and thinnest bodies and the smallest limbs had the advantage over their fellows that they were able to crawl through narrower holes, and that for this reason a greater number of them survived to produce offspring. Here we have a better basis for reasoning than the supporters of Lamarck's doctrine, because we actually know that longer parents, in whom this quality was apparently not the result of taking thought, do tend to produce on the average longer offspring.

3. The view of the mutationists, already foreshadowed by Aristotle, and in recent years especially associated with the names of Bateson and de Vries, expresses the conclusion that the evolution of new species has taken place principally by the help of variations of the discontinuous kind. By this process there can arise at a single step new forms which have already the complete and definite character usually associated with a species specially adapted to particular conditions. Of these new forms, those which happen to be fitted for their surroundings as well as or better

2

than their predecessors will survive, whilst those which are worse will be destroyed by the action of natural selection.

Thus it would be an appropriate use of this conception to seek in a mutation the explanation of the final loss of the much reduced limbs presumably exhibited by comparatively recent ancestors of the family of snakes. This final loss is especially difficult to understand on the Darwinian theory. Moreover, changes of a closely similar nature are not hypothetical, but have actually been observed to take place. At the same time it must not be supposed that mutations are confined to the loss of pre-existing organs ; indeed, the origin of a totally new organ is quite inexplicable on either of the two preceding theories. The very first inception of such an organ must, it would seem, of necessity be sudden.

After giving some account of the earlier theories of evolution, we shall next proceed to treat of those subjects with which we are more properly concerned— that is to say, the recent experimental observations on variation and natural inheritance, together with their bearing on the theories of evolution. And in the first place we shall describe some recent studies which are not strictly experimental, but which nevertheless deal to some extent with actual facts—namely, the statistical study of variations, particularly of continuous variations. This subject has been dignified by a special name, and is now described as the science of biometry.

Of even greater interest, however, are the more strictly experimental researches which have been published within the last five or six years. In the first place, we have the observations of de Vries, who has introduced a new method of study—that of cultivating great numbers of seedling plants with the object of discovering definite new forms or mutations among their number. Lastly, and in its results much the most important of all, we have the method of Mendel, published half a century ago, but only recently brought into prominence owing to its rediscovery and confirmation by three independent workers —Correns, Tschermak, and de Vries. This method consists in the cross breeding of strains of plants or animals which differ in definite characters, and in the statistical examination of the proportions in which these characters appear among the offspring obtained from the crosses.

Further experiments on the lines which Mendel indicated bid fair to revolutionize within a few years the arts of the breeders of plants and animals. This is due to the fact that such experiments are leading to the introduction into these pursuits of a degree of scientific exactness which was previously altogether unforeseen. The change in our ideas regarding the method of hereditary transmission of characters, which has resulted from these experiments, has been aptly compared with the change brought about in men's understanding of the science of chemistry by Dalton's conception of the atom. For the rest the new experiments tend on the whole to confirm the experience of practical breeders ; only the elucidation of one simple rule of inheritance has brought into order a host of phenomena, which were previously quite incapable of a coherent explanation.

The experimental results with which it is the province of this book to deal are, then, firstly those of biometry, or the statistical study of variations, and particularly of continuous variations ; secondly, the results of direct observations bearing upon the origin of species by the discontinuous method ; and thirdly, the results of experimental observations on heredity by the methods of scientific breeding. By these methods results of the utmost moment to mankind have been, and are being, arrived at, quite apart from their interest as bearing upon the problems of evolution. From a biologist's point of view, however, the latter is, of course, paramount. And so it has been thought fitting to begin with a brief discussion of the problems of evolution, and of the various solutions of them which have been from time to time suggested.

In a later chapter some of the more prominent recent results of the kindred science of cytology—the microscopic study of the minute constituent parts of organisms—will be briefly described, on account of the very close connection which recent progress in this subject bears to the experimental study of the inheritance of the grosser characters.

CHAPTER II

EVOLUTION may be defined as progress involving differentiation, an ever-growing complication of things which accompanies almost all the operations of Nature. The idea of a differentiation of this kind may be enforced by a homely and quite imaginary illustration of such a process. Imagine the proper ingredients of a plum cake to be very finely minced and intimately mixed together, so as to form a more or less homogeneous material. Then, if by any means the separate particles of currants, raisins, peel, and so forth, could be made to segregate out in such a way as to give rise to the ordinary structure of this pleasant confection, we should have arrived at the structure of a plum cake by a process of evolution involving considerable differentiation.

The progressive increase in complexity which is characteristic of so many natural processes is in great part occasioned by the fact that a single ' cause ' is followed as a general rule by more than one ' effect.' This apparently simple circumstance was pointed out by Herbert Spencer,* who has perhaps done more than any other to establish and emphasize the general applicability of the evolution idea. For the law of origin by evolution is by no means exclusively confined to the method of coming into existence of the species of animals and plants. On the contrary, it was equally well applied by Spencer himself to describe the manner in which are supposed to have arisen the stars and other heavenly bodies, the geological strata and geographical configuration of the earth, and the various gradations of human society.

The discovery that certain chemical elements exist which are themselves not immutable has been made since Spencer's time. Quite recently ' the phenomena of radio-activity have forced us to believe that radium is passing continuously into helium,'† and something more than a suspicion has been aroused that radium is itself derived from uranium. Thus the dreams of the alchemists are shown to have been not wholly without foundation, for the probability is strong

* Spencer gives the following illustration : Regarding the striking together of two bodies as a ' cause,' he points to the following possible ' effects ' : A sound ; other vibrations or movements in the surrounding air ; a disarrangement of the particles of the two bodies in the neighbourhood of the point of collision ; the production of heat, and possibly of a spark— *i.e.*, of light.

Two words in this sentence are placed between inverted commas, to indicate that they are used in a strictly popular sense. The use of the words ' cause ' and ' effect,' though seldom strictly scientific, is often convenient, and if used with caution, there is no reason why they should lead to misunderstanding. See Whetham, ' The Recent Development of Physical Science,' chapter i.

† Whetham.

that under suitable conditions other matter may be
observed to behave in the same way as radium. More
than this, Professor J. J. Thomson has been able to
describe the atoms of the elements as different aggrega-
tions of a single kind of corpuscles, and to show that
a progressive change in the number of corpuscles
making up the atom is accompanied by a progressive
alteration in the properties of the atom itself, so that
it has now become possible to establish a theory of
the evolution of the chemical elements themselves.

Passing from the almost immeasurably small to the
almost immeasurably great, we may briefly consider
the probable mode of origin of the solar system from
an extremely diffuse cloud of material substance, according to the famous nebular hypothesis of Laplace.
By a long-continued process of contraction under the
influence of gravity the nebular substance came to be
of varying density, and acquired a rotary movement
in one plane. As the mass continued to contract
owing to the mutual attraction of its particles, the
velocity of rotation increased, until at last the increas-
ingly rapid motion of the outermost ring of the now
lens-shaped nebula gave rise to a centrifugal force great
enough to counteract the tendency to contraction, and
in the further condensation of the mass this ring was
left behind. The ring next broke down at one point,
and contracting on itself gave rise to a single spheroidal
body which acquired a movement of rotation in the
same direction as that of the parent nebula. This
body was the outermost planet Neptune, and the rest
of the planets were produced in a similar manner,

until at last a central mass was left, and this became
the sun. Satellites were thrown off from several of
the planets just in the same way as the planets themselves arose from the original nebula, and Saturn's
rings are pointed to as showing this process even now
in course of operation.

Such a description as this may appear fanciful at
first sight, but it was worked out quantitatively as
well as qualitatively by its author, and was shown to
explain in detail a multitude of phenomena. Spencer
points out that when we have, worked out by one of
the first of mathematicians, a definite theory of plane-
tary evolution based on established mechanical laws,
and one which accounts in a satisfactory way for all
the known phenomena, the conclusion that the solar
system really did arise by a process of evolution is, to
say the least, difficult to avoid.

The establishment and propagation of the idea that
the present condition of the earth's surface arose
through a course of gradual evolution, by the agency
of such processes only as are known to be in operation
at the present day, is the great contribution of Sir
Charles Lyell to the science of geology. We may
briefly trace the evolution of the idea itself, beginning
with the speculations of Werner, who, from observations of the geological formations of a limited tract of
country, came to the conclusion that the successive
strata were precipitated one by one from an universal
ocean. Here we see the first germ of the idea of
evolution embodied in the notion that the stratified

rocks came into existence gradually and through the operation of a supposed natural cause.

A great advance upon Werner's theory was made by Hutton, who, observing the formation of strata at the present day from the sediment washed down by rivers, concluded that the ancient strata were deposited in the same manner. Since, by the long continuation of this process the continents must gradually become reduced to the level of the sea, Hutton supposed that at long intervals of time the action of subterranean heat came into play, and fresh continents were upheaved, a process accompanied by the outpouring of the igneous rocks, the true origin of which he had duly recognised. In this theory a hypothetical cause still survives, since we have no actual experience of vast upheavals of the kind which Hutton supposed to have taken place. Lyell showed that such slight changes of level as are known to be in progress at the present day, especially in association with the phenomena of earthquakes, might, if continued over a long series of ages, give rise to the necessary amounts of elevation. Lyell also pointed out a number of subsidiary causes of disintegration and deposition of strata of the kind which can still be seen in operation at different parts of the earth's surface. At the present time it is sometimes thought that Lyell went a little too far in his championship of the cause of uniformity. Lyell supposed that the agencies which may now be everywhere observed in operation, such as rain and rivers, the sea, volcanoes and earthquakes, were sufficient to account for all the phenomena which the crust of the earth exhibits. It is now more generally supposed that in very early times forces similar in kind to those in action at the present day may have exhibited considerably greater violence.

To produce the present condition of the surface of the earth by the action, gradually accumulated, of such processes of denudation and upheaval as are now going on around us, vast periods of time are clearly necessary. The early evolutionists, having once got rid of the idea that the date given by Bishop Usher as that of the creation of the world is a necessary and integral part of religion, immediately allowed their imaginations to run riot with regard to the amount of time at their disposal. Since this question of the extent of geological time has an important bearing on the problem of organic as well as upon that of inorganic evolution, it will be well to pay some attention to more recent views upon the subject.

Some years ago the generous ideas of biologists as well as of geologists were to a great extent shattered by the calculations of Lord Kelvin. These were based upon three separate sets of data, which we may enumerate without entering into a lengthy explanation of the calculations involved. The evidence made use of consisted of (1) the rate of the earth's rotation, as affected by tidal retardation ; (2) the rate of secular cooling of the earth, as deduced from the rate at which the temperature of the earth's crust rises on passing inward from the surface ; and (3) the rate of cooling of the sun by radiation. The three calculations were found to show a very fair measure of agreement, and they led to the conclusion that considerably less than a hundred million years has elapsed since the first formation of seas upon this planet, an event which must have preceded the possibility of aqueous geological action and the existence of living organisms.

Allowing for the circumstance that geological processes may have gone forward with considerably greater rapidity during the earlier periods of the earth's history than is the case at the present day, the time thus allowed by the physicist is generally regarded by geologists as too little. Reckoning from the known rate of denudation, which is, of course, the same as the rate at which the same material is deposited beneath the sea, Geikie, who admitted, however, that such data are only of a very rough description, concluded that the space of a hundred million years would afford sufficient time for the laying down of the known aqueous strata. But there can be little doubt that the lower metamorphosed rocks represent a much longer period of time than the primary, secondary, and tertiary epochs added together ; consequently, the respective estimates of Lord Kelvin and the geologists appear to be contradictory. The recent discovery of the enormous quantities of energy stored up in radio-active substances introduces a serious modification into the mathematical argument from astronomical data, and Sir George Darwin ' sees no reason for doubting the possibility of augmenting the estimates of solar heat, as derived from the theory of gravitation, by some such factor as ten or twenty,' on the supposition that a considerable proportion of the sun's substance was made up of radio-active material.

The above remarks may serve to illustrate the importance of the theory of evolution as applied to the two sciences of astronomy and geology. We pass next to a brief historical consideration of the development of the evolution theory as a method of describing the origin of the species of animals and plants.

The views of the ancient Greeks cannot be said to have much more than a purely speculative interest. Some rudiments of the idea of evolution have been attributed to Empedocles as well as to several other early writers, and in the writings of Aristotle, for whom the too great faith of his successors for many ages has been followed by a somewhat unmerited degree of contempt in modern times, we find that the evolution idea had reached quite a respectable degree of development.

In the Middle Ages the adoption of the Jewish cosmogony by the Christian Churches effectually annihilated all useful thought upon the subject of species, since the hypothesis of separate creation affords no scope for further speculation or experiment, and it is not until the end of the seventeenth century that we find thoughtful men beginning to struggle against the ecclesiastical bondage. Thus Erasmus Darwin derived the idea of generation rather than creation of the world from David Hume, and himself waxes enthusiastic over the thought :

' That is, it (the world) might have been gradually

produced from very small beginnings, increasing by the activity of its inherent principles, rather than by a sudden evolution of the whole by the almighty fire. What a magnificent idea of the infinite power of THE GREAT ARCHITECT ! The Cause of causes. Parent of parents. *Ens entium.*'

De Maillet, writing in 1735, showed a definite idea of the production of existing species by the modification of their predecessors. At the beginning of the nineteenth century similar speculations were published by Goethe and by Treviranus, and the latter was the first to apply the term ' biology ' to the science of the phenomena of life. Lamarck about the same time provided a definite theory as to the method by which the modification of species takes place.

Before discussing Lamarck's hypothesis and the alternative theories more recently proposed, it will be well to pass in review the evidence upon which is based our belief that the species of animals and plants have arisen through the modification of pre-existing species, and to show that the greater part of this evidence is quite independent of any views which we may adopt as to the actual method by which a particular species came into existence. And in the first place we may point out the entire absence of any evidence, direct or indirect, in favour of the alternative supposition of a special creation of each separate species.

The evidence for evolution falls naturally into a number of fairly well defined sections :*

* A modification of the list given by Huxley, ' Collected Essays,' vol. ii., p. 205.

1. THE GRADATION OF ORGANISMS.—Both in the animal and vegetable kingdoms we may trace, in spite of certain gaps, a long series of gradations in complexity of structure, so that between the simplest and the most complicated of living things a great number of intermediate stages are to be found. When we pass to the lower end of the scale in either case, we come upon a group of creatures of comparatively simple organization. Among them we find members with regard to which we cannot definitely say that they are either animals or plants. Moreover, these unicellular organisms resemble in many ways the egg-cell from which every individual among the higher animals and plants originates.

2. EMBRYOLOGY.—All the members of a particular group of animals or plants as a rule resemble one another more closely in the early stages of their individual development than they do in the adult condition, and in the earliest stages of all they are often indistinguishable. These facts are explained if we suppose that such individuals have a common origin, that they are descended from a common ancestor, and that traces of their pedigree are still to be observed in the developmental stages through which each one passes. We do not find a complete parallelism between the development of the individual and the history of the race, nor should we expect to do so, since embryonic as well as adult stages may be modified in the course of evolution; what we should expect is a more or less vague historical sketch, and this is what is usually found remaining.

3. MORPHOLOGY.—On comparing together the different members of one of the great groups or classes of animals or plants, we find the same fundamental plan of organization running through all of them. Series of corresponding organs are often to be made out which are built upon the same general scheme, although their functions may be quite dissimilar; so that, for instance, in the hand of a man, the paw of a dog, the wing of a bat, and the paddle of a whale, almost identically the same series of bones can be traced. An obvious explanation is to be found in the supposition that these parts have arisen by the divergent modification of parts which were originally identical.

4. GEOGRAPHICAL DISTRIBUTION. — Observation shows that groups of closely allied creatures are often found living in neighbouring districts, and that when such a barrier as an ocean or a range of lofty mountains is passed an entirely new fauna and flora are usually to be met with. These facts may be explained by the hypothesis that allied groups of species originated by a process of descent in the same countries which they now inhabit, and they can be explained by no other known hypothesis.

5. THE GEOLOGICAL SUCCESSION OF ORGANISMS.— The general facts regarding the distribution of allied species of animals and plants in time point in precisely the same direction as those relating to their distribution in space. In a few cases, notably in that of the extinct horse of North America, a long chain of possibly ancestral types has been found leading back

to a remote and very different progenitor. This supposed ancestor of the horse was a creature little larger than a moderate-sized dog. It had four separate toes to each fore-limb, and three to each hind-limb, and its teeth were much simpler and less specialized than those of existing horses. The general distribution of organisms throughout the geological strata agrees, moreover, in a remarkable way with what is to be expected on the evolution theory.

6. CHANGES UNDER DOMESTICATION.—Among domesticated animals and plants we know of numerous cases in which the actual origin of new forms has been observed. These have often differed from their predecessors by amounts quite comparable with the differences by which natural species or even genera are separated. A notable example of this process is afforded by the numerous breeds of pigeons known to have arisen under domestication from a single wild species. We have no reason whatever for supposing that domesticated species are more mutable than wild species, and there is consequently every reason to believe that changes of a similar character take place in Nature. The conditions of domestication, of course, afford much better opportunities of observing such phenomena.

7. THE OBSERVED FACTS OF MUTATION.—Nevertheless, individual specimens of particular wild species are frequently found showing modifications which, if they occurred constantly in an isolated group, would afford a basis for the description of new species. In a few cases the actual occurrence of similar changes has been observed in wild species of plants.

We see, therefore, that the evidence in favour of the existing species of animals and plants, having arisen by a process of evolution, is of a most ample and convincing kind. The theory of organic evolution is, however, incomplete until we have arrived at a true account of the method or methods by which new species arise from old ones. The earliest definite explanation, as already stated, was that given by Lamarck, and we may next proceed to consider the Lamarckian theory of the origin of species.

Earlier writers had already supposed that species became modified through the action of the external conditions to which they were exposed. Lamarck laid special stress upon the observed facts that the organs of individuals become increased and developed through use, and that disuse is followed by a dwindling and loss of the power of action. By the inherited effects of use and disuse, and of modifications caused by external conditions, Lamarck supposed all evolution of species to have come about.

Reference has already been made to Lamarck's description of the method of origin of the characteristic form of snakes, owing to the endeavours of the snakes' ancestors to creep through narrow passages. Lamarck was quite consistent inasmuch as he explained the different types which have arisen among domesticated species by the same theory as he applied to the origin of species in a state of nature. Thus he supposed the differences between race-horses and heavy cart-horses to be the direct result of the different kinds of enforced exercise to which the ancestors of these races were respectively subjected. Similarly, all the different breeds of dogs were supposed to have arisen owing to the different habits which the various successors of the first domesticated dogs acquired, small changes being accumulated by inheritance in each successive generation.

Turning now to species in a state of nature, the case of the giraffe is one of those most often quoted. Lamarck supposed a comparatively short - necked ancestor of the giraffes to have taken up the habit of browsing upon the leaves of trees, owing to the difficulty of obtaining other food in an arid region. In order to obtain their new food the animals were obliged to be continually stretching upward, and the effort to elongate their necks was attended with some small measure of success in each individual. This increase, being accumulated by inheritance in every succeeding generation, ultimately led to the great stature exhibited by the giraffes of the present day.

The stilt-like legs of many wading birds were ascribed by Lamarck to the result of the continued attempts of ancestors which had shorter extremities to obtain their food in shallow water without wetting their feathers. The long-continued endeavours of these birds to stretch and elongate their legs had the same effect as the similar efforts made by the ancestors of the giraffes. It has been suggested, however, by a critic of Lamarck's position that such birds would be likely to eschew fish dinners long before any notable increase in the length of their legs was arrived at.

If some of the above cases appear a little ludicrous,

there are other instances in which the Lamarckian explanation seems adequate, and where an alternative hypothesis is lacking. Such a case is afforded by the family of the flat fishes, including such well-known species as the sole and plaice. In the adult condition these fishes lie flat on one side ; and during their development from the young condition, that eye which, if it remained in its original position, would look directly downwards travels round the head until it comes to lie quite upon the upper surface. As Darwin pointed out, agreeing in this with Mivart, a sudden spontaneous transformation in the position of the eye is hardly conceivable, and it is equally impossible to explain the origin of this remarkable feature by the action of natural selection, because a slight change in the position of the eye could be of no advantage so long as this organ remained upon the under surface. The very young fish, whilst still symmetrical, are known sometimes to fall upon one side, and when in this position to twist the lower eye forcibly upwards. Darwin himself therefore supposed that the origin of the adult structure is to be attributed to the inherited effect of efforts of this kind.

The interest of the last case lies in the fact that it relates to a structure, the origin of which does not appear explicable on the theory of natural selection ; its bearing will therefore be better understood when we come to discuss that theory in the next chapter.

The inherited effects of voluntary striving can clearly have no application to the case of plants. Lamarck therefore supposed that evolution in the vegetable kingdom had taken place entirely through the action of external agencies upon plants. The soil, for example, in which a plant grows has a direct influence upon its form. Altitude, moisture, heat, and light are other important factors, and the effect of their influence upon the plant was supposed by Lamarck to be inherited. The shape of irregular flowers was regarded as having been directly caused by the strains and pressures occasioned by bees and other insects whilst making their visits in search of honey or pollen.

Lamarck's theory turns entirely upon the question whether acquired characters are inherited, and if so, to what extent, since, if such inheritance is shown to be extremely slight, the cause, though a true one, may be insufficient to explain the effects attributed to it. Now, theories' of heredity apart, and leaving aside the results of minute observations which had not been made in Lamarck's time, the natural supposition undoubtedly is that acquired characters are inherited just as much as any others. Given the observed fact that offspring resemble their parents more closely than they do other members of the same species, it is natural to believe that the child will take after the forms exhibited by its parents at the time of its conception rather than after those shown by them at any previous period of their lives. This seems to be the natural view in the absence of any other evidence for or against, and so accurate a thinker as Herbert Spencer, writing before the publication of the ' Origin of Species,' regarded the term inheritance as necessarily implying inheritance of this particular kind.

For this reason it has sometimes been thought that Darwin scarcely accorded to Lamarck the appreciation which he deserved ; and yet Darwin himself fell back upon the Lamarckian explanation on the few occasions when natural selection seemed to have failed him.

When, however, we come to know more of the actual facts of sexual generation, we find that it is very difficult, if not impossible, to imagine any kind of mechanism by which this supposed transmission of acquired modifications can take place. We shall defer the further discussion of this subject, as well as the question of the existence of direct and other evidence of use inheritance, until the latter half of the next chapter, where we shall refer briefly to the controversy upon these subjects which followed the establishment of the principle of natural selection.

CHAPTER III

IN 1813 a communication was read before the Royal Society by Dr. W. C. Wells upon the differentiation which exists between certain races of mankind. In Dr. Wells's paper this differentiation was explained from the facts that, since no two individuals are alike, some would be better fitted than others to resist the diseases proper to a particular country, and would consequently tend to survive, whilst their less fortunate neighbours would perish in greater numbers. Wells supposed the dark races of mankind to be better adapted to warm climates than white races are, and he thus applied to the particular case of the human species the true Darwinian principle of a gradual evolution through the survival of the fittest.

A similar view was applied to the origin of species in general by Patrick Matthews in a book on naval timber and arboriculture published in 1831.

Both these works were unknown to Darwin at the time of the first publication of the ' Origin of Species,' and it is quite unnecessary to point out that their existence does not in the least prejudice the value or originality of that great work. Their interest at the

38

present time is merely historical, as showing the direction in which thought was tending in the earlier half of the nineteenth century.

Before the ' Origin of Species ' was published, A. R. Wallace communicated to Darwin a paper in which the bearing of the same idea was worked out at some length. and this paper was read, together with an abstract of Darwin's own views, at a meeting of the Linnean Society in July, 1858.

With this notice of other claimants to the idea of natural selection we may proceed to give an account of the theory as it is developed in the earlier chapters of the ' Origin of Species.'

We must first glance at Darwin's method of using the term variation. Darwin applied this term to every kind of difference which is found to occur between parents and their offspring, or between members of the same family, no matter whether these differences were great or small. It has since been shown that a number of quite distinct phenomena were in this way regarded from a single standpoint, without a proper discrimination being made between them. But the differences between continuous and discontinuous variation, quantitative and qualitative variation, and the rest, were not pointed out until long subsequent to 1859. Thus, beyond recognising a distinction between sports and individual differences, and attaching greater weight to the latter kind of changes, as being those which chiefly led to the origin of new species, Darwin made no further analysis of the facts of variation, but accepted all sorts of differences between individuals as affording the material upon which natural selection might be expected to operate.

The idea that a selective influence exists in Nature arose from a study of the remarkable effects produced in the case of domestic animals and plants by the action of *artificial* selection. Darwin seems, however, to have been a little credulous in accepting the statements of certain breeders with regard to their power of producing any desired new type to order. Now that scientific men are themselves beginning to make experiments in breeding, with the check of exact records to act as a drag upon the exuberance of the imagination, they are becoming somewhat sceptical as to the mystic and almost miraculous powers attributed to the old-fashioned breeders, though, indeed, Mr. Luther Burbank would seem to be a survival from the period we speak of, if the statements of his recent enthusiastic biographer are to be credited.* Less gifted but more methodical observers find that they have no creative powers worth speaking of, and that all they can do is to keep a sharp look-out for the novelties which Nature may send them.

Selection, whether natural or artificial, can indeed of itself have no power in the direction of creating anything new ; its influence is destructive or preservative, but nothing more than this. The breeder keeps the new forms which take his fancy, and destroys the rest ; that is the whole story.

* Harwood, ' New Creations in Plant Life.' Mr. Burbank certainly seems to have a really wonderful instinct for the discovery of curious and useful novelties.

Yet a remarkable number of new kinds of creatures are known to have arisen in this way, and their diversity is no less astonishing, as a visit to any great show of domestic plants or animals will at once demonstrate. Here may be seen varieties of pigeons, for example, like the carrier, pouter, fantail, and tumbler, which, if they were found existing in a wild condition, would be placed in separate genera by any ornithologist. The domestic races of fowls, dogs, horses, sheep, and cattle show scarcely less divergence, and modifications no less remarkable have been perpetuated in the case of many cultivated species of plants. Whilst these types have survived, being deliberately preserved on account of their use or beauty or curious appearance, a still greater number have doubtless been exterminated, either because they did not attract the breeder's favourable attention, or on account of their having passed out of fashion.

Darwin sought in Nature a substitute for the baleful judgment of the breeder, and found it in an extension of the Malthusian doctrine to organic beings in general. The idea which is identified with this expression did not, however, originate with Malthus, nor does that author claim it as his own, as the following extract from the first chapter of the ' Essay on Population ' will show :

' It is observed by Dr. Franklin that there is no bound to the prolific nature of animals and plants but what is made by their crowding and interfering with each other's means of subsistence. Were the face of the earth, he says, vacant of other plants, it might be gradually sowed and overspread with one kind only, as, for instance, with fennel ; and were it empty of other inhabitants, it might in a few ages be replenished from one nation only, as, for instance, with Englishmen.'

Malthus' 'Essay' was first published in 1798, and was subsequently much enlarged. Its author proved incontrovertibly, by a survey of facts gathered from almost all the countries of the world, that human population tends to increase in a geometrical ratio, and that, consequently, after a time, the less gifted classes of any community are bound to suffer from a stress of poverty, only partly relieved by a high infant mortality, periodic famines, and similar factors, or in less civilized countries by infanticide and other artificial checks.

Among animals and plants in a state of nature the rate of increase is often very much greater than in the case of the human family, and even where it is not so, unchecked breeding would in a comparatively few years lead to the overpeopling of the earth with the descendants of a single pair. As an example of the rate of increase shown by a wild species, we may consider the case of the elephant, instanced by Darwin himself, since this is regarded as being one of the slowest breeders among all known animals. Darwin assumes that the elephant begins breeding at thirty years, and continues to do so until it reaches the age of ninety, bringing forth six young in the interval, and surviving to the age of a hundred. Then, if there were no casualties, he calculates that after from 740 to 750 years there would be nearly nineteen million elephants alive descended from the first pair.

Let us also consider the case of a minute rapidly breeding animal of a typical kind. My friend Mr. R. C. Punnett has recently been engaged upon an experiment which involved the breeding of rotifers, a kind of animal barely visible to the naked eye. They were bred for sixty-seven generations, and each individual produced on the average thirty eggs. The whole experiment occupied less than a year, yet Mr. Punnett calculated that if he had been able to rear all the animals which, at this rate of breeding for this number of generations, were theoretically obtainable, he would have become the possessor of a solid sphere of organic material with a radius greater than the probable limits of the known universe.

This geometrical rate of increase is common in a greater or less degree to all living organisms. Since the space and food-supply available for the support of any species has no corresponding tendency to increase, it follows that a large proportion of the individuals born must perish before they reach the adult state, or at least without producing offspring. Darwin's contention is that there will be a strong tendency for those individuals which show slight modifications in the direction of a better adaptation to their environment to survive at the expense of those of their brethren which do not exhibit similar modifications. This is the principle called natural selection by Darwin, and by Herbert Spencer the survival of the fittest. Let us quote Darwin's own summary of the process :

' If under changing conditions of life organic beings

present individual differences in almost every part of their structure, and this cannot be disputed ; if there be, owing to their geometrical rate of increase, a severe struggle for life at some age, season, or year, and this certainly cannot be disputed ; then, considering the infinite complexity of the relations of all organic beings to each other and to their conditions of life, causing an infinite diversity in structure, constitution, and habits, to be advantageous to them, it would be a most extraordinary fact if no variations had ever occurred useful to each being's own welfare, in the same manner as so many variations have occurred useful to man. But if variations useful to any organic being ever do occur, assuredly individuals thus characterized will have the best chance of being preserved in the struggle for life ; and from the strong principle of inheritance, these will tend to produce offspring similarly characterized. This principle of preservation, or the survival of the fittest, I have called Natural Selection. It leads to the improvement of each creature in relation to its organic and inorganic conditions of life, and, consequently, in most cases, to what must be regarded as an advance in organisation. Nevertheless, low and simple forms will long endure, if well fitted for their simple conditions of life.'*

We have here a very definite and concise statement of the way in which Darwin believed the principle of natural selection to take effect in the production of new kinds of organisms. It will be our business in this and in succeeding chapters to show how far the modern

* ' Origin of Species,' sixth edition, p 96.

study of the nature of individual differences and of other kinds of variations, as well as of the manner of operation of ' the strong principle of inheritance,' has confirmed this view as to the method of origin of species, or has led to the introduction of modifications.

Let it be remembered that this suggestion of a natural means of modification had, within a few years, the effect of convincing practically the whole thinking world of the truth of the theory of organic evolution— an effect which all the other arguments recited in the last chapter were quite unable to produce, so strong was the then existing prejudice in favour of the doctrine of special creation.

The truth of the general principle of the survival of the fittest is quite untouched by recent criticism ; but a great deal of argument has been expended over the questions : (1) how much fitness is sufficient to lead to survival, and (2) whether very small advantages in the way of fitness, even if they lead to the survival of the individuals which exhibit them, will be followed to an indefinite extent by further improvements in the same direction in succeeding generations. We shall find that a good deal of evidence has accumulated tending to show that the second of these questions must be answered in the negative, although the point is not yet settled to the satisfaction of everyone.   The remainder of the present chapter will be occupied in discussing some of the arguments which bear upon this question.

The fact that organic beings on the whole are, as a general rule, very closely fitted for the conditions in which they have to pass their lives is clearly shown by the study of adaptations. This is a subject which those followers of Darwin who believe in the all-sufficiency of natural selection have brought into considerable prominence. For a full account of many supposed beautiful adaptations, from the point of view of the most prominent member of the school in question, reference may be made to Weismann's recently published book, ' The Evolution Theory.'

On the theory of natural selection in its extreme form, all the parts of an animal or plant—or, at any rate, all the points in which one species differs from another nearly related species—are supposed to have arisen on account of their usefulness to the creatures possessing them. Every detail of structure is thus regarded as being more or less closely adapted to the circumstances which attend the life of the animal or plant in question. This adaptation is never, indeed, regarded as perfect, because natural selection is always in progress, and its work is never absolutely done; but the point is that the features of every part are aimed at some useful purpose ; or, if they are not, then they have been useful in former times and under different circumstances, and are now undergoing a process of gradual removal, because the individuals in which the useless structure is least developed will now have the best chance of surviving. That the form and structure of an animal or plant is in general closely fitted to its environment is of course true ; otherwise the creature would very soon cease to cumber the earth. But the student of adaptation goes into details, and endeavours to find a use for every minute point of structure, on the assumption, which we shall presently see to be open to criticism, that but for their usefulness these details would not exist. We may proceed to glance at one or two examples of the kind of thing which is meant when it is said that an animal or plant exhibits very marked adaptative features.

The family of whales belongs to the class of mammals of which the more typical members are land animals possessing four legs, and having their bodies covered with hair. We may deal in particular with the Greenland or true whalebone whales, since these are in many ways the most specialized members of the group.

The Greenland whale* has a spindle-shaped body like that of a fish, and its fore limbs are modified into flippers resembling the pectoral fins of fishes. The hind legs are only represented by a few rudimentary bones, which are completely hidden within the body wall, and the function of propulsion, which is performed by the hind legs in such less completely aquatic animals as seals, is here taken over by a great tail-fin which resembles that of a fish, except that it is placed horizontally. Hair is absent, but under the skin a thick layer of blubber is developed, which prevents a too rapid loss of heat, and at the same time adjusts the specific gravity of the body to that of the surrounding water. External ears are entirely wanting, and the waves of sound are apparently transmitted to the drum of the ear directly through the bones of the head.

* Weismann, The Evolution Theory,' English edition, ii. 313.

The external openings of the nostrils are placed quite on the upper surface of the head, so that the animal can breathe whilst almost completely submerged ; and the larynx is so modified that the function of swallowing does not interfere with that of breathing. Perhaps the most remarkable feature of all is the enormous development of the head, and especially of the mouth. The huge jaws, in combination with the extraordinary plates of whalebone which fringe the edges of the mouth and act as a sieve, enable the animal to get its nutriment from the minute free-swimming creatures with which the surface waters of the ocean abound. Associated with this special method of feeding is the fact that teeth are only to be recognised in the embryo, and afterwards entirely disappear.

The whales differ in all these points from any other mammals, and failing almost any of these differences, they would not be able to live in the special conditions in which they find themselves. It must therefore be admitted that we have here a case of very close adaptation of an animal to its natural surroundings, and one which extends to almost every detail of its structure. Darwin himself, moreover, has been at special pains to show how some of the most remarkable of these structural adaptations may possibly have arisen through natural selection.

One of the most remarkable cases of mutual adaptation, in which an animal and a plant are associated together, is shown by the method of fertilization observed to take place in the flowers of the Yucca plant of the Southern United States. The act of pollination is performed by a moth—*Pronuba*—which possesses special organs particularly adapted for this purpose, in the shape of peculiar maxillary tentacles which are found in no other kind of moth. The female has also a long ovipositor with which she can pierce the tissues of the ovary of the plant, and so lay her eggs within it. With the aid of her peculiar tentacles the female moth collects from several flowers a ball of pollen of considerable size, which she kneeds into a firm pellet. She then carries this to a different flower, and after depositing a few eggs in the ovary she climbs to the top of the style and presses the ball of pollen into the stigma. Thus the ovules of the flower are fertilized, and whilst some are eaten by the larvæ of the moth, others develop into seeds and reproduce the plant.

The foregoing are perhaps two of the most remarkable cases known of animals having peculiar habits, and possessing at the same time special organs which render them well fitted for these habits and no others; but many other cases of scarcely less wonderful adaptations have been pointed out.

Darwin himself indicated the direction in which the study of adaptation was to proceed, and his books on ‘Insectivorous Plants’ and on the ‘Fertilization of Orchids’ afford us a delightful insight into a number of adaptive contrivances which are to be seen in plants. Another very interesting series of adaptive characters are those which have been gathered together under the heads of Protective Resemblance and Mimicry, and these have a special interest for us. because they illustrate the way in which the zeal of the seeker after adaptive contrivances may run away with him if not kept well in hand. For there is scarcely any limit to the number of problematical cases which have been described as adaptive resemblances, and so explained as having arisen through natural selection, whilst the evidence in favour of such a supposition is in many cases highly questionable. On the other hand, in a number of well-marked instances, the theory of mimicry certainly seems to afford an adequate explanation of the way in which many curious characters and structures may possibly have come into existence.

The family of the mantises, including the walking-stick and leaf insects, affords many examples of animals which both in their colour and configuration show a very close resemblance to surrounding inanimate objects. This resemblance must have the effect of concealing them from their enemies, and more particularly from their prey, as, indeed, a study of their habits indicates quite clearly.

Phyllopteryx, an Australian fish allied to the well-known sea-horse (*Hippocampus*), is provided with a number of irregular appendages of ragged skin resembling the seaweed amongst which this animal is found. In this way the characteristic symmetrical appearance of a live animal is got rid of, and the creature is rendered extremely difficult of observation. Here, again, the concealment afforded is probably useful in leading to the deception of the smaller fishes upon which the creature feeds.

Examples of this kind in which the shape of an animal leads to its concealment are comparatively infrequent, although a considerable number might be collected. On the other hand, some resemblance between the *colour* of an animal and its surroundings is to be traced in the majority of the members of many groups. Familiar examples are afforded by the white colour of animals which live in snow, the tawny grey colour of most desert species, the green of grass-frequenting animals, and so on. It is perhaps not quite certain that in some of these cases the peculiar colour is not evoked by the direct action of some cause which affects different species in the same way; but such a cause awaits discovery, and in the meantime natural selection has certainly a strong claim to be regarded as the proper explanation.

A more strict use of the term mimicry, however, is to restrict it to cases where one species apes the colour pattern or other external character proper to another species which inhabits the same region; and the idea of mimicry has been put forward as especially appropriate in cases where the mimicked species is common, and can be thought to possess some special means of protection. Several supposed examples of this phenomenon have been described in the case of different genera of tropical butterflies, but the best of them seem to be open to criticism, since there is nothing to prove that colour patterns of the same type may not have arisen from the same causes in quite different groups. In cases where the environment to which the different forms were exposed was similar—as would be the case in a single locality—such a process of parallel evolution might be thought to be all the more likely.

Resemblances can only be properly explained as representing cases of mimicry when both the species concerned—the mimic and the mimicked—inhabit the same locality ; but plenty of cases of matching between the colour patterns of insects which live in quite different parts of the world could also be pointed out. Let us take a concrete example. Everyone is familiar with the flower-frequenting flies (*Syrphidæ*) which are to be seen hovering about plants in sunny weather. These insects closely mimic the appearance of various small bees and wasps, the habits of which are similar. Here, then, is surely a case where the deceptive resemblance to an animal well armed in its sting must cause prospective enemies to let these flies alone. In Southern Japan, as Dr. Andreæ pointed out to me, flies of this kind are surprisingly numerous, and their resemblance to bees particularly noticeable. So abundant are they that, from the point of view of the flowers which they visit, these flies doubtless provide an efficient substitute for the bees of other countries, which are here conspicuous by their absence. But if real stinging insects are wanting, or even very scarce, the supposed enemies of the flies can have no experience of the ill-effects produced by catching them. How, then, can these flies benefit from their resemblance to bees ?

This kind of thing must make us somewhat suspicious of supposed cases of mimicry even between species possessing the same range.

When the ideas of mimicry and protective resemblance are carried into the vegetable kingdom, as they have been by some writers, absurdities are soon found to arise. Thus it has been suggested that the leaves of dead nettles resemble those of the common nettle for the sake of the protection so afforded, and that the mottled stems of certain tropical herbaceous plants gain a similar immunity on account of their resemblance to snakes.

In plants a great number of fanciful resemblances between different species can be detected, and some between plants and animals, very few of which can be supposed to be of any possible utility to the species which exhibit them. They must be regarded as cases of parallel evolution, the causes of which are quite unknown. Such resemblances as that between the shoots of *Casuarina indica* and those of the common horse-tail, between *Saxifraga hypnoides* and certain mosses, between the horse- and Spanish-chestnut, between the seed of a pine and the fruit of an ash-tree, are so frequent in the vegetable kingdom as to be the delight of malicious examiners in elementary botany. It is impossible to believe that in such cases the resemblance is in itself of any value to either species, and few people will be found to maintain that the likeness of a bee- or spider-orchis to an insect is of any utility to either animal or plant.

But if resemblances can arise which are useless, and which, consequently, cannot be explained through natural selection, it becomes uncertain whether this principle can hold good as the true description of the origin of any sort of resemblance. On the other hand, resemblances which are useful will tend to *survive* through natural selection in whatever way they may have arisen. This last consideration will account for the frequency with which apparently adaptive likenesses are to be found in nature, even if we suppose that their origin was 'accidental,' or simply due to the operation of similar external causes. The same criticism applies to all cases of adaptation of whatever kind, so far as concerns their supposed method of origin by the action of natural selection upon individual differences.

Perhaps a still more serious criticism of the methods of those who spend their time in seeking out or devising cases of adaptation has been made by Bateson. who points out the logical difficulty that we can never make any quantitative estimate of the amount of benefit or the reverse which any particular structure may afford to its possessor. It is easy enough to imagine particular circumstances in which an organ developed in a particular way may be of undoubted service, but whether the net amount of such service throughout the life of the creature considered is greater or less than the strain upon its resources caused by the development of such an organ is quite beyond our powers of determination.

' The students of adaptation forget that even on the strictest application of the theory of selection it is unnecessary to suppose that every part an animal has, and everything which it does, is useful and for its good. We, animals, live not only by virtue of, but also in spite of what we are. It is obvious from inspection that any instinct or organ *may* be of use ; the real question we have to consider is *how much* use it is.

To know that the presence of a certain organ *may* lead to the preservation of a race is useless if we cannot tell how much preservation it can effect, how many individuals it can save that would otherwise be lost ; unless we know also the degree to which its presence is harmful ; unless, in fact, we know how its presence affects the profit and loss account of the organism.'*

A great many other criticisms and objections have been brought at various times against the theory of the origin of adaptations by the action of natural selection, and many of these were considered and replied to by Darwin in the later editions of the 'Origin of Species.' We shall only consider here a few which have been put forward more or less recently. Before doing so it will be well to point out once more that no one questions the validity of natural selection as a means of exterminating types which are unfitted for their environment—there is clearly a tendency for the fittest types to survive once they have come into existence. Nor can there be any doubt that species in general are well adapted to the conditions which their environments present. But when this is admitted it does not necessarily follow that natural selection, directing the accumulation of minute differences, has been the method by which these adapted forms have originated.

The power of regenerating a lost part must clearly often be of service to the creatures which possess it. Such a power may in many cases be considered to be a well-marked adaptation. But, as Morgan has well

* W. Bateson, ' Materials for the Study of Variation,' p. 12.

pointed out, there are insuperable difficulties in the way of adopting the belief that such a power can have been acquired through the action of natural selection. Let us consider the power of regenerating a lost tail, such as is shown by the common gecko, or wall-lizard, of the tropics. To account for this power by natural selection we should have to suppose, firstly, that every stage in the growth of a partly regenerated tail, even its first small rudiment, was useful to the animal ; and, secondly, that there was so much competition between lizards which had lost their tails, that those which could regenerate them a little better would survive rather than the others. The first of these suppositions as to the utility of a partly regenerated tail is in the highest degree improbable ; but against the second there is an entirely fatal argument, since, if the lizards which regenerated badly were exterminated owing to competition with those which had better powers of regeneration, much more would all the injured lizards be exterminated in competition with those which had escaped injury.

The theory of sexual selection constitutes an important branch of the Darwinian account of the origin of specific structures. We are here concerned with this hypothesis only in so far as it leads to a criticism of the efficacy of natural selection from another point of view. By the theory of sexual selection Darwin attempted to explain the origin of two sorts of characters in particular, one or other of which frequently appears in the male sex only of many of the higher animals.

In the first place, we have to notice the presence of special weapons, such as horns or tusks, developed exclusively or to a special extent in the males of those species in which it is the habit of the members of this sex to strive together for the possession of the females. In such cases the stronger and better-armed males are supposed to survive, and to leave a greater number of offspring than their weaker rivals, so that this form of competition is regarded as acting in quite a similar way to natural selection.

In a second set of cases, of which many remarkable instances are to be seen among birds, the males are found to exhibit brilliant and varied colours, or to possess special decorations in the form of plumes or other appendages, or to be gifted with the power of song. It is to cases such as these that the term sexual selection more properly applies, since the females are supposed to bestow their favours upon the most beautiful males, and to reject the advances of those among their suitors which are less lavishly provided with ornament.

In these cases, where the development of brilliant colours or other ornamental arrangements is believed to have taken place owing to the choice of the females —particularly in such a case as is represented by the peacock's tail or the wings of the Argus pheasant— the supposed change must have come about in direct opposition to the action of natural selection, since the latter would favour a production of colours resembling those of the natural environment for the sake of concealment, and would hinder the formation of such exaggerated appendages on account of the loss of activity which they must entail. We are, therefore, obliged to conclude that natural selection is much less rigorous in its action than some people have supposed, for if this principle is inadequate to prevent such an exuberance of form and colour in these particular cases. its action becomes open to question in other cases as well.

Similarly, Morgan finds a difficulty in understanding why natural selection has not led to the extermination of species which are handicapped by the existence of internecine strife between the males, in favour of other species which faced the battle of life with united strength. But in this argument it seems to be forgotten that examples of the kind of strife in question are most frequent among herbivorous animals, where the struggle for existence must be chiefly determined by the quantity of vegetable food which the individuals can obtain, so that the loss of the weaker males may not be a disadvantage. Moreover, Darwin's conclusion that natural selection is most rigorous between members of the same species is left out of account.

The preceding arguments seem to show that in particular cases certain structures and phenomena associated with species cannot be explained as having arisen through the unaided action of natural selection. When weighed against the great mass of evidence which Darwin accumulated in support of his theory, these few considerations cannot be said to be in any way fatal to the belief that natural selection of minute differences has played an important part in the origin of species. Still, they add in some measure to the weight of recent evidence which points to the conclusion that many specific structures have had a different method of origin. We have already pointed out that there are two alternative methods, each of which has its adherents. Before passing to a consideration of the now prevalent view of mutation, something still remains to be said with regard to the remaining theory—the theory of Lamarck.

Darwin himself, as we have seen, admitted the minor importance of the inheritance of acquired characters, as well as that of the phenomenon of sporting, regarding both these processes as causes of the origin of new species subsidiary to the action of natural selection upon individual differences, whilst he looked upon the latter as the main process in organic evolution.

Later writers, however, have asserted that natural selection is the sole cause of the origin of species, and in particular they have denied any effect to the inheritance of acquired characters—the Lamarckian factor—asserting that there is not, and cannot be, any such inheritance. Among the most distinguished opponents of the theory of use-inheritance were A. R. Wallace, the co-discoverer of natural selection ; and Professor Weismann, who has argued the case with particular ability. Much the most able defender of the principle of use-inheritance was Herbert Spencer, who was one of the few who had thoroughly convinced themselves of the truth of the theory of evolution years before the ' Origin of Species ' made its appearance. Since all arguments in favour of the evolution of species were incomplete unless some means by which such an evolution could take place had been suggested, Spencer adopted the Lamarckian theory of modification, and to this he always firmly adhered, though admitting the validity of natural selection as an additional factor in the process. Some of Spencer's arguments in favour of a belief in the inheritance of acquired characters are well worth repetition, since they have never been altogether refuted.

Herbert Spencer's argument consisted mainly in the enumeration of structures the origin of which cannot be explained by natural selection. On the other hand, the inheritance of acquired characters, if this form of inheritance could be proved to have a real existence —as Spencer believed it could—was shown to be a perfectly adequate explanation of the origin of the structures in question. In 1893, when Spencer upheld his opinion for the last time, Bateson had not yet pointed out that the facts of definite and discontinuous variation afford an alternative way out of some of these difficulties. In the absence, therefore, of any other effective cause, the result of the argument pointed strongly to the conclusion that the inheritance of acquired characters must be a reality.

The first of Spencer's arguments was based upon the different powers of tactual discrimination which are to be found in different parts of the human body. The degree of this sensitiveness may be estimated by the use of a pair of compasses, the points of which can be set at different distances apart. It is then found that with the tip of the forefinger the points can be distinctly recognised as two when they are separated by no more than $\frac{1}{12}$ inch. When applied to the middle of the back, on the other hand, the points must be opened to a distance of $2\frac{1}{2}$ inches before the sensation of a single touch becomes resolved into two distinct sensations.

The distribution of this power of discrimination over the surface of the body is approximately as follows :

Tip of tongue $\;..\qquad..\qquad..$	$\frac{1}{24}$ inch.
Tip of finger $..\qquad..\qquad..\qquad..$	$\frac{1}{12}$,,
Inner surface of second joint of	
$\quad$ finger $\;..\qquad..\qquad..\qquad..$	$\frac{1}{6}$,,
Tip of nose $\;..\qquad..\qquad..\qquad..$	$\frac{1}{4}$,,
Cheek, palm of hand, and end of	
$\quad$ great-toe $\;..\qquad..\qquad..$	$\frac{1}{2}$,,
Forehead $\;..\qquad..\qquad..\qquad..$	$\frac{3}{4}$,,
Back of hand, crown of head $\quad..$	1 ,,
Breast $\;..\qquad..\qquad..\qquad..$	$1\frac{1}{2}$,,
Middle of back, middle of thigh,	
$\quad$ middle of forearm $\;..\qquad..$	$2\frac{1}{2}$,,

Now, it is out of the question to suppose that natural selection can account for all these differences. An increased sensitiveness of the tips of the fingers might, indeed, be of so much use as to give the individual possessing it a definitely increased chance of survival. But it is hard to believe that it can be important for a man to have the tip of his tongue twice as sensitive as the tips of his fingers. And why should the tip of the nose be more sensitive than the cheek, or the cheek than the top of the head, or the breast than the back ? In the last case it might even be suggested that in a savage, since the sense of touch is the only one with which his back is provided, it might be useful for that surface to have acquired a more delicate sense of touch than the anterior surface, which is guarded by the power of vision, as well as being more readily explored by the sensitive finger-tips. If such an argument is regarded as far-fetched, so in an equal degree must be any attempt to explain the actually observed distribution through the action of natural selection.

On the other hand, Spencer points out that the series of parts enumerated in the above table stands in almost exactly the order of the frequency with which the members composing it are actually exposed to tactual experience.

The tongue is perpetually in contact with the minute unevennesses afforded by the surfaces of the teeth.

The palm of the hand and the lower joints of the fingers are used chiefly in grasping, and not in the more minute manipulations for which the finger-tips are employed. And the experience of the back of the hand in coming into contact with various irregular bodies is not to be compared with that of the palmar surface, yet it is very much greater than that of so unexposed a part as the middle of the forearm.

For the carrying on of his argument. Herbert Spencer has shown that increased use of the power of discriminating small objects by touch is accompanied by an increased degree of sensitiveness in individuals. Blind people use their finger-tips in this way to a much greater extent than those whose sight is unimpaired.

Two blind boys examined by Spencer were both found to be able to distinguish with the tips of their fingers points separated by only $\frac{1}{14}$ inch. And two skilled compositors could both distinguish in this way points placed no more than $\frac{1}{17}$ inch apart, so that a person with a trained sense of touch acquires a considerably finer development of this faculty than an ordinary individual.

If, then, acquired characters of this kind are inherited, even to an extremely minute extent. such as would be scarcely perceptible in a single generation, the account of the origin of the observed phenomena would be complete.

As a second argument, Herbert Spencer points out the difficulty of accounting for the development of co-ordinated sets of structures by the action of natural selection upon separate minute variations of the several parts concerned.

The enormous horns of the ancient Irish elk, weighing in some cases over a hundredweight, required specially strong neck muscles, bones, and ligaments, and strong fore legs for their support. But an increase in the strength of a single muscle following increased weight of the horns would be useless if unaccompanied by a corresponding increase in many other structures, and, if useless, could not be selected. The chance of all the parts concerned varying simultaneously in a corresponding direction is very small if these variations are really independent, and the chance of their doing so repeatedly is in such a case infinitesimal.

Let us take another case of a similar nature. The

hind legs of such an animal as a cat are admirably adapted for the purpose of making a spring. In order to arrive at such a structure by the modification of limbs previously adapted only for running, changes must occur in almost all the bones, muscles, and ligaments of the limbs, and these changes must keep pace with one another so that one part may not grow out of proportion with the rest. It is quite impossible to suppose that this can be effected by the natural selection of minute fortuitous variations of the various parts, each occurring independently. But simultaneously with these changes the fore legs have become modified in a totally opposite direction. They have become straight, firm, and pillar-like for receiving the weight of the body in the downward leap. Compare, says Herbert Spencer, the silence of a cat's leap up on to a table with the thud made by the fore legs as it jumps down upon the floor.

Modification of the fore legs and of the hind must thus have proceeded in almost exactly opposite directions in the two cases, and in each a great number of parts are separately co-ordinated. For natural selection to have had any effect, all the co-ordinated parts of one pair of legs must have varied in one direction, whilst similar parts in the other pair of legs varied simultaneously in another direction. It is out of the question to suppose that this could have happened simply by chance.

' What, then, is the only defensible interpretation ? If such modifications of structure produced by modifications of function as we see take place in each individual are in any measure transmissible to descendants, then all these co-adaptations, from the simplest up to the most complex, are accounted for. In some cases this inheritance of acquired characters suffices by itself to explain the facts; and in other cases it suffices when taken in combination with the selection of favourable variations. An example of the first class is afforded by the change just considered; and an example of the second class is furnished by the case, before named, of development in a deer's horns. If, by some extra massiveness spontaneously arising, or by the formation of an additional " point," an advantage is gained either for attack or defence, then, if the increased muscularity and strengthened character of the neck and thorax, which wielding of these somewhat heavier horns produces, are in a greater or less degree inherited, and in several successive generations are by this process brought up to the required extra strength, it becomes possible and advantageous for a further increase in the horns to take place, and a further increase in the apparatus for wielding them, and so on continuously. By such processes only in which each part gains strength in proportion to function can co-operative parts be kept in adjustment, and be re-adjusted to meet new requirements. Close contemplation of the facts impresses me more strongly than ever with the two alternatives—either there has been inheritance of acquired characters, or there has been no evolution.'*

As we pointed out in the last chapter, there seems at

* Herbert Spencer, 'The Inadequacy of Natural Selection,' p. 29.

first sight to be no inherent difficulty in the way of acquired characters being inherited. Weismann has, however, pointed out a very serious difficulty, which is brought into prominence on making a study of the minute anatomy of the cells of organisms during the earlier stages of their development.

In the ordinary course of events every one of the higher animals and plants begins its existence in the form of a single minute cell—the fertilized ovum or egg. This cell exhibits no trace of the complicated series of organs which will develop from it when it is subjected to the proper conditions. When the egg is placed in favourable circumstances with regard to warmth, moisture, food-supply, and the like. it first divides into two equal portions; and microscopic study shows that elaborate precautions are taken to insure the equal bipartition of its minute constituent parts. Each of the two cells thus formed divides again into two further cells, and by a series of repeated bipartitions of this kind the cells which constitute the adult body are at last brought into existence. Since the body soon becomes differentiated into a number of unlike organs, it is clear that at certain stages of the process the two cells arising from a division must come to differ slightly from one another; and the cells ultimately produced show very considerable differences of form, structure, and size. Among all the cells which finally arise those which have undergone the least modification from their original condition are those from which are developed the sexual reproductive cells, or germ-cells, of the organism. Indeed, Weismann concludes that there is no reason for supposing that these have undergone any modification at all.

If we consider the cells which build up an adult organ, and for the moment regard each separate cell as an individual, we see that each of these individuals possesses an ancestry of cells stretching right back to the fertilized ovum—the single cell in which the whole organism originated. So far as the later cell divisions are concerned, the cell-lineage of a particular organ is separate and distinct from that of the cells of any other organ. At a certain distance back in the history of the organism we shall come across a common cell-ancestor for the cells belonging to a pair of neighbouring organs and the more widely separate the parts to which the cells we are considering belong the further back must we go before we find their ancestry merging in a single cell. In a similar way as with other organs, so it is found that the sexual cells or germ-cells of an adult organism have a history quite distinct from that of the cells of any other part of the body; and these cells are the only ones which are concerned in the formation of the offspring. Thus we see that the particular cell-lineage leading up to the germ-cells is the only one which is continued into another generation; all the others terminate with the death of the individual creature of which they form a part. From this point of view we may consider the nature of a given series of animals as being determined only by the particular series of cells which constitute the direct ancestry of the germ-cells in each individual; the cells which make up the bodily structure are the result of so many offshoots which come to an end at the death of the organism, and leave no progeny of their own.

Wilson has expressed this view of Weismann's very clearly : ' It is a reversal of the true point of view to regard inheritance as taking place from the body of the parent to that of the child. The child inherits from the parent *germ-cell*, not from the parent body, and the germ-cell owes its characteristics not to the body which bears it but to its descent from a pre-existing germ-cell of the same kind. Thus the body is as it were an offshoot from the germ-cell. As far as inheritance is concerned the body is merely the carrier of the germ-cells which are held in trust for coming generations.' (The diagram illustrating Weismann's theory of inheritance is a modification of that given by Wilson.*)

In the light of this conception it may be seen that the idea of the inheritance of a modification acquired by an adult bodily organ is comparable with the supposition that if a man develops his muscles by exercise his brother's children will be thereby modified.

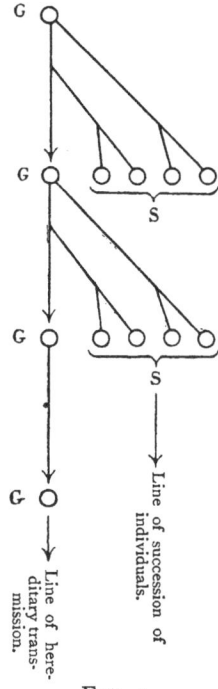

FIG. 1.

G, Germ cells ; S, Somatic cells.

* ' The Cell in Development and Inheritance,' p. 13.

The minute study of the germ-cells, taken in connection with modern experimental work on the methods by which inheritance takes place, shows a strong tendency to confirm Weismann's view, so far as the case of distinct and definite characters is concerned. But if we regard such definite characters as having arisen by definite steps or mutations according to the view now gaining ground, the study of them will have no bearing upon the question of use-inheritance, since use does not lead to large and definite changes in the individual, but to comparatively small changes of a quantitative kind.

There are some, including de Vries, who regard all fluctuating variations (individual differences) as being of the nature of acquired characters, and as being at the same time capable of hereditary transmission, although de Vries believes the amount of progress possible in this way to be strictly limited. Let us see if there is any way in which a transmission of such characters can be conceived of.

It must be pointed out that the cells which make up an organism are not completely marked off and separated from one another ; on the contrary, it seems impossible to doubt that reactions may take place between them long after their first formation. Indeed, Sedgwick has shown that in a number of diverse kinds of animals there is never any sharply limiting barrier between cells at all, and this writer has gone so far as to speak of animals in general as being built up of a continuous network of protoplasm with nulcei at the nodes. In plants, too, though at first sight their constituent cells seem to be cut off from one another like so many closed boxes, it has been shown that there is almost universal communication between the pioto-plasmic masses so enclosed, in the shape of minute fibrils of living substance which traverse the interven-ing walls.

It would thus seem possible for liquid or easily soluble substances to pass freely from one part of the body of an organism to another. It is possible, for example, supposing the enlargement and strengthening due to the exercise of a particular muscle to be associ-ated with an increased production of some definite chemical substance, to imagine that an increased amount of the same substance might become enclosed in the germ-cells, so that this substance would be present in the offspring in greater abundance than would have been the case if the muscle of the parent had not been exercised. And this might render more easy a further development of the muscle by exercise in the next generation. In a similiar way increased bulk following upon better nutrition might be inherited, and this de Vries seems to have succeeded in showing to be actually the case in plants. Such changes might normally be so slight as to be almost imperceptible in a few generations, and yet after many generations might accumulate to an important extent. It would be impossible in practice to distinguish changes of this kind from what are known as accidental individual differences, and, indeed, there is no evidence at hand to disprove de Vries' assertion that all continuous variations are of the nature of acquired characters—

and we know that continuous variations are inherited.

On the other hand, several lines of inquiry have separately led to the conclusion that a great number of the visible characteristics of organisms are of a definite kind, and are inherited definitely, their appearance being determined by the presence of definite structures or substances in the germ-cells. The evidence, as we shall see later on, points to the conclusion that such characters have arisen suddenly at a single step, and we must conclude that in such a case a definite change in the germinal structure has been followed by a definite alteration in the character of the organism arising from the germ ; since no one can suppose that a large and definite structural alteration can be first acquired by the adult organism and then inherited by its offspring—such a process is unthinkable.

Thus we see that the inheritance of acquired characters, if such inheritance really takes place at all, must be confined to the transmission of changes of an indefinite and quantitative kind—to the case, in fact, of continuous variations or individual differences. Moreover, there is nothing to show that all continuous variations are not of the nature of acquired characters.*

* We know, at any rate, that continuous variations are not invariably due to the cause which Weismann supposed—namely, to the mingling together of characters derived from the two parents—a supposition which is of fundamental importance to his theory—because, as Karl Pearson has pointed out in this connection, parthenogenetically reproduced organisms, in which no such mingling has taken place, may be just as variable as those which owe their origin to the process of sexual generation.

It is possible that variations of this nature may gradually lead to important and even to specific changes, but whether this is the case still remains to be proved. On the other hand, we shall see that specific differences do sometimes arise at a·single step, and there is strong but indirect evidence to show that this is the way in which a very great number of specific differences have actually arisen. Indeed, some have contended that this is the universal process by which such differences originate, but this again is not proved, nor is it altogether likely. In any case the inheritance of acquired characters can have nothing to do with that of definite and discontinuous differences.

This is a problem to which we shall return in the concluding chapter, in the light of further evidence concerning continuous and discontinuous variations and their manner of inheritance, which will be by that time available.

FRANCIS GALTON.

[*To face p.* 73.

CHAPTER IV

IN the present chapter we have to consider in some detail the manner in which purely statistical methods have been applied to certain biological data, a proceeding to which the term biometry has been attached by Professor Karl Pearson. Before concluding our account we shall give a brief sketch of some of the more important evidence bearing upon the problems of evolution which has been brought to light by the methods of biometrical science.

The first investigator to apply the methods of statistics to the solution of biological problems was the Belgian astronomer, Quetelet. In 1845, in the form of a series of letters addressed to the Grand Duke of Saxe-Coburg and Gotha, Quetelet published an admirable account of the theory of probability and its relation to human affairs, and one in which the use of advanced mathematics was avoided. The pioneer of biometry in this country is Francis Galton, whose book on ' Natural Inheritance ' embodies an extremely lucid introduction to the statistical study of variation and inheritance. From these two works are derived most of the ideas submitted in the present chapter.

The more recent advances in biometry are mostly the result of work published by Professor Karl Pearson ; they consist largely in the elaboration of mathematical methods of dealing with statistical problems, and as such it would be inappropriate to give any further account of them here.

The mention of the word ' statistics ' at once raises a certain prejudice in the ordinary mind ; in common parlance, the unreliability of arguments based upon statistics is sometimes treated as proverbial, and as used in biology they have, as a matter of fact, one very serious danger at least. Statistics deal with groups and not with individuals, and there is a real difficulty involved in the fact that the average of a group may represent something quite different from any individual which the group contains, whilst at the same time a group may include individuals of very diverse natures. Nevertheless, when used without prejudice to the future examination of individual inheritance by more detailed investigations, the methods of biometry have undoubtedly yielded information of great value to the evolutionist, particularly in the case of such material as that afforded by the human race, since the application of precise experiments to this particular species is at present out of the question.

Some students of biometry, however, would go very much further than this, for it is their professed opinion that their own form of study is the only method by which any real advance in our understanding of the processes of evolution can be brought about. This opinion is based upon the assumption, of which proof is wanting, that new species have arisen exclusively through the accumulation by natural selection of variations of a strictly indefinite, fluctuating, or normal kind. We have already seen reasons for believing that this is very far from being the case, and future chapters will be found to add considerably to the force and quantity of the evidence already adduced.

Normal variations, strictly speaking, are individual differences which can be supposed to depend upon a large number of small factors or causes—factors so numerous and so minute that the numerical distribution of the individuals examined, when ranged in order according to the feature chosen for examination, is found to conform closely to that which would be expected on the mathematical theory of chance. Such a distribution will only result when the differences considered can be strictly regarded as lying upon a linear scale, and when they are also evenly distributed along that scale. That is to say, the biometrician deals with continuous variations of a quantitative kind. It is to be hoped that these somewhat obscure sayings will be more easily understood in the light of what follows.

The facts of variation have not been found readily amenable to precise definition, but we shall endeavour to make plain by the aid of a few examples what kinds of variations do and what kinds do not appear to be legitimate objects for the application of biometrical methods. Thus it may be thought that the biometrician is outrunning his license when he ranks the colours shown by the iris of the human eye in a continuous series of eight shades, because in doing so he groups together a number of probably definite factors with others which are of an indefinite kind. When the colours of the human eye come to be studied in greater detail, there can be little doubt that they will be found to depend upon some such factors (among others) as the following :

1. (*a*) Definite differences in structure, and (*b*) the definite presence and absence of pigment in certain definite positions ; as well as—

2. (*a*) Indefinite variations (individual differences) in structure, and (*b*) in quantity of pigment—if, indeed, the quantitative differences are not found to be also definite.

In the above example a suitable and legitimate object for biometrical investigation would be the differences in amount of a particular pigment.

But definite differences may also exist in the case of an apparently simple quantitative character. The accompanying figure (Fig. 2) shows the variations in length of the fruits of three different but closely allied species of evening primrose, as measured by de Vries. In this diagram the vertical distances are in each case proportional to the number of individuals having particular lengths of fruit, and the actual length of the fruit is in each case proportional to the horizontal distance from an imaginary vertical line some way to the left of the figure ; the points thus plotted are joined by straight lines, so that a polygonal figure is obtained representing the nature of the variation in each particular case. The diagram shows at once that the species A and C have each a characteristic mean size of fruit, and the existence of this definite mean is not affected by the fact that the range of variation

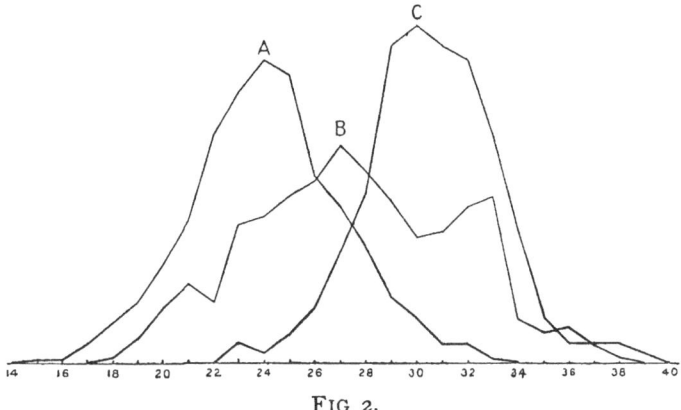

FIG 2.

overlaps in all three cases. Species B, on the other hand, seems to show signs of division into at least two separate groups.

Differences of a similar kind are sometimes to be found among the progeny of the same individuals. Races of garden peas may be selected which, amongst other differences, are characterized by the presence of large and of small seeds respectively. In each case there is variation of a normal kind about a mean value, but in each case the mean is quite distinct. There is evidence that if a race of large-seeded peas is crossed with a small-seeded variety, and the resulting cross-bred plants are self-fertilized, their progeny in the second generation will be separable into different groups, and some of these will show almost exactly the same size characteristics as those which were exhibited by the two original parental strains. The only difficulty in the way of invariably distinguishing the two original kinds, after their segregation in the offspring of the cross, lies in the fact that the smallest seeds of the large type may be smaller than the largest seeds of the smaller strain, and this is a difficulty which applies equally to the original strains before crossing, as well as to the case of the evening primrose fruits just mentioned.

Now, it is clear that if we mixed together the seeds of several different races of peas in the proper proportions, the result might lead to a normal distribution of the kind presently to be described. The several races, however, would none the less be perfectly distinct, even though we could not separate the individual seeds belonging to each by any direct method.* Such a mixture of races would constitute a decided pitfall for the unwary statistician, and it is well to remember that, after even the most elaborate mathematical analysis. the final result cannot be clothed with any greater amount of certainty than the facts from which the calculation set out. Those who have made a large expenditure of intellectual effort in such processes have, unfortunately, a natural tendency to overlook this elementary fact.

Prior to the application of statistical methods to a particular case of normal variation a number of preliminary processes have to be gone through

* It would generally be possible to decide which strain a particular seed belonged to by sowing it and observing the variation of its offspring.

Having selected a particular character for investigation, we must make a quantitative estimate of its development in each member of a fair sample of individuals which show the character in question. What is to be understood as a fair sample was well expressed by Quetelet when he wrote that statistics must be collected without any preconceived ideas, and without neglecting any numbers. We shall find that in this point the biometrical method differs from the method introduced by Mendel, since in the latter careful discrimination of data is an essential feature.

The quantitative determination of a character may be made either by counting or by measurement. That is to say, we must proceed by measurement if the character we are dealing with is one of size or weight, and by counting if the character shows a series of numerical values of its own—*e.g.*, if it is such a character as the number of veins in a leaf or the number of stigmatic bands on a poppy capsule. Before we make any determinations we ought to be quite certain that we are dealing with the same character in each individual, and that the individuals themselves are truly comparable with one another. Thus we might make a series of measurements of a particular bone in a particular limb of a particular race of human beings with some assurance that we should be dealing with homogeneous material.

Our measurements or countings will fall either naturally or artificially into groups. In the case of countings the groups are naturally limited by the

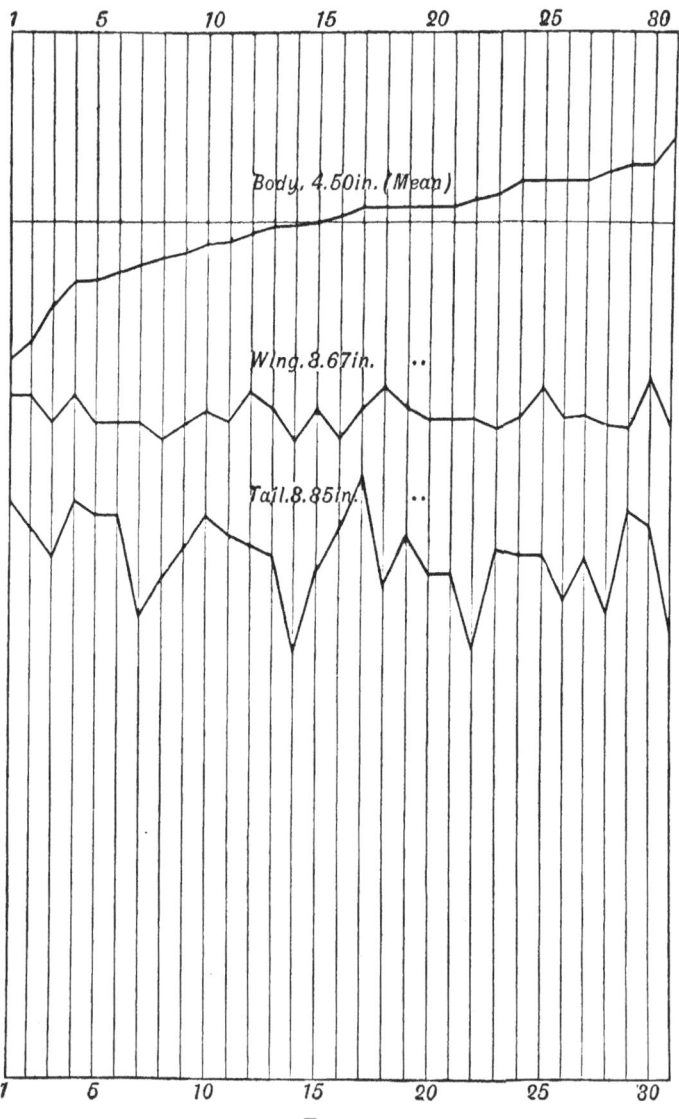

FIG. 3.

numbers which represent the character of each individual, whilst measurements are artificially limited through the fact that they have to be made in units of some kind—*e.g.*, to the nearest inch or some other value. Such groups, characterized by equality of range—each, that is to say, covering an equal number of units—are technically known as *classes*.

Thus if we are dealing with human stature, and if our measurements were made only to the nearest inch, all the individuals of 6 feet in height would fall into one class, those of 6 feet 1 inch into another class, and so on. If, on the other hand, we were engaged in counting the number of ray florets in the heads of daisies, a class would include all those heads which possessed a particular number of rays.

Without division into classes, however, a survey of a comparatively small number of measurements may be facilitated by ranging the values in some kind of order. This is done, for example, in the accompanying figure for the measurements to hundredths of an inch of the lengths of the body, wing, and tail of thirty-one specimens of a North American bird. The diagram is taken from A. R. Wallace's ' Darwinism.'

Even with this small number of measurements the diagram brings out two points very clearly. In the first place, there is no close correspondence between the variations in length of body, wing, and tail. Secondly, in the case of body-length, in respect of which the specimens are ranged in order, the number of individuals of a medium size is seen to be greater than the number of those which show extreme values. This

6

excess of mediocre individuals comes much more prominently into view as soon as a larger number of measurements can be considered, and the results arranged in a different way.

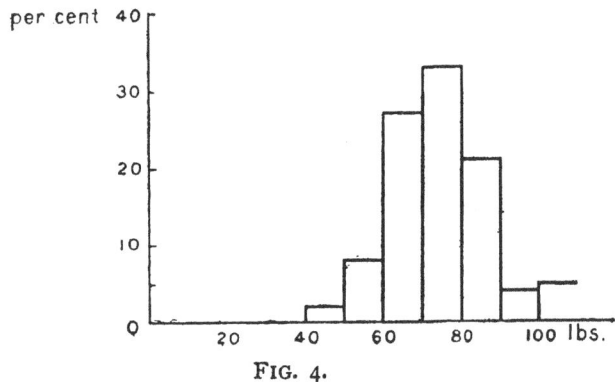

FIG. 4.

The above diagram is constructed from the entries in the third column of the accompanying table, which is taken from Galton's 'Natural Inheritance.' It repre-·sents the variations in the strength of pull (as exerted by an archer in drawing a bow) shown by 519 men as recorded at the International Health Exhibition in 1884. Here equal distances measured off along the base line represent equal increments in the strength of pull of the right hand, and the vertical heights of the rectangles erected upon these bases represent the percentage numbers of the men examined which exhibited each value of the character under consideration. In this example it is easy to see that the central class is the largest, whilst the extreme classes contain a comparatively small number of individuals.

TABLE I. (FROM GALTON).

STRENGTH OF PULL (519 MALES, AGED 23-26).

From Records made at the International Health Exhibition
in 1884.

Strength of Pull.	Number of Cases observed.	Percentages.	
		Number of Cases observed.	Sums from beginning.
Under 50 lbs.	10	2	2
,, 60 ,,	42	8	10
,, 70 ,,	140	27	37
,, 80 ,,	168	33	70
,, 90 ,,	113	21	91
,, 100 ,,	22	4	95
Above 100 ,,	24	5	100
Total	519	100	

Finally, we may display in a somewhat more detailed
fashion the result of a still larger number of measure-

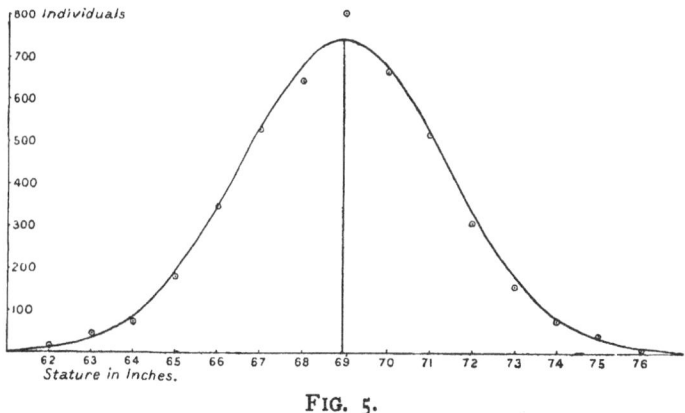

FIG. 5.

ments.   Fig. 5 shows the variation in stature of a large
number of members of Cambridge University of British

6—2

extraction, and exhibits in a concise form the result of 4,426 measurements recorded by the Cambridge Anthropometric Society. In this figure the stature in inches is indicated on the base line, whilst the perpendicular distances indicate the number of cases in which each particular height was recorded. The separate classes in this case include those who were found to fall within the limits of $\frac{1}{2}$ inch on either side of each consecutive integral inch of stature, measurements which fell exactly half-way between two classes—*e.g.*, one of 69$\frac{1}{2}$ inches—being reckoned as a half to each of the classes in question. The continuous line in the diagram represents the form of the ' normal curve ' which approximates most nearly to the line obtained by joining together the points actually plotted.

There seems to be good evidence that in such a case as that of human stature the figure obtained in this way will approximate more and more closely to the shape of what is known as a normal curve, according as the number of individuals measured and the accuracy of the measurements increase.

In order to arrive at a proper understanding of this fact, we must consider the derivation of the ' normal ' curve from another point of view—namely, from the point of view of the mathematical theory of probability, which it will be our endeavour to present in as simple a manner as possible.

Let us consider the result of tossing up a number of similar coins simultaneously. If we toss up two coins only we may get any of the following results : (1) Head head, (2) head tail, (3) tail head, (4) tail tail.

And it is clear that any one of these combinations is equally likely to appear on any given occasion, if the coins are supposed to be strictly symmetrical, and are tossed up entirely at random. Now, the second and third results are the same unless the two coins are individually distinguishable. So we may write the most likely result of tossing up two pennies four times in the following way :

1 H H + 2 H T + 1 T T.

And in a similar way we may discover that the most likely result of tossing up three coins eight times is :

1 H H H + 3 H H T + 3 H T T + 1 T T T.

In the first case H T is twice as likely to appear as H H at any single throw, and in the second case H H T is three times as likely as H H H in any single toss.

It is possible to work out the most probable relative frequency of the various possible combinations in the case of any number of coins. Thus for ten coins the sequence of numbers runs :

TABLE II.

Heads.	Tails.	Relative Probability.
10	0	1
9	1	10
8	2	45
7	3	120
6	4	210
5	5	252
4	6	210
3	7	120
2	8	45
1	9	10
0	10	1

These values are plotted in the accompanying diagram (Fig. 6) as vertical distances above a base line. The figure obtained by joining together the points thus arrived at may be observed to show some resemblance to the previous Figs. 4 and 5.

The three series of numbers already given are those which are obtained on expanding the expressions $(1+1)^2$, $(1+1)^3$, $(1+1)^{10}$. In general the probabilities of the various possible combinations when n coins are tossed simultaneously are given by the expanded value of $(1+1)^n$.

Quetelet has worked out the relative probabilities of the most frequent combinations in the case of 999 coins simultaneously tossed—*i.e.*, the expanded value of $(1+1)^{999}$. A few of these values are given in the following table :

TABLE III.

Heads.	Tails.	Relative Probability.
500	499	1·0
501	498	0·996
502	497	0·988
503	496	0·976
504	495	0·961
505	494	0·942
510	489	0·803
520	479	0·432
530	469	0·155
540	459	0·037
550	449	0·006
560	439	0·0006

It may thus be seen that the likelihood that a result appearing in any given throw will show a still greater

difference in the relative number of heads and tails
than 560 : 439 becomes very small indeed. Although

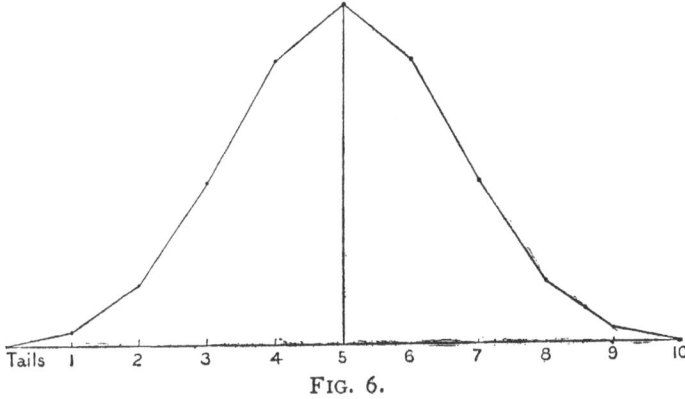

Tails 1 2 3 4 5 6 7 8 9 10

FIG. 6.

a throw of all heads or all tails is possible, the odds
against such a result being ever actually seen are

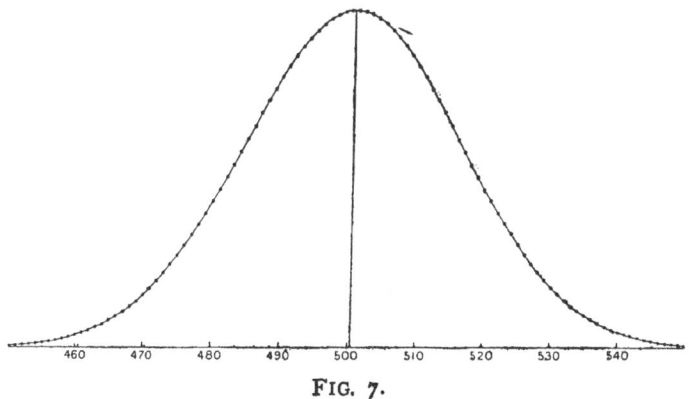

460 470 480 490 500 510 520 530 540

FIG. 7.

almost inconceivably great. In Fig. 7 the middle
values for $(1+1)^{999}$ are plotted—those combinations

being included which lie between 550 : 449 and 449 : 550. The points thus obtained are so close together that the eye can scarcely distinguish whether they are joined by straight or curved lines. We have, in fact, arrived at a close approximation to the normal curve.

The curve thus approximately indicated may be seen to be closely similar to the one shown in Fig. 5 ; in fact, the two curves are of such a kind that by altering the vertical and horizontal scales in one of the figures in a suitable ratio their form could be made practically identical.

The figure arrived at in this way approximates to a mathematical curve which is intelligible to the mathematician from the formula $y = e^{-x^2}$. The theoretical curve is really arrived at by supposing n in the expression $(1 + 1)^n$ to become indefinitely great. Practically, by making n very large we can get as near an approximation as we may wish to the normal curve of theory. Even in the case of relatively small values of n the approximation to the normal curve is fairly close, as may be seen by comparing together Figs. 6 and 7.

The example of tossing up coins was only taken as a means of illustrating the more general assumption of an event or a magnitude depending upon a number of causes of equal strength, which in the long-run act with equal frequency in two opposite directions. We can understand that human stature may afford a comparable case, when we consider the large number of bones and cartilages the lengths of which must be added together in order to make up the total stature of any individual, and that the separate length of each one of these elements depends upon factors which we have no means of classifying exactly.

It now becomes necessary to mention one or two technical terms which are used in connection with the normal curve. The *mode* of such a curve is the longest perpendicular which can be drawn from the

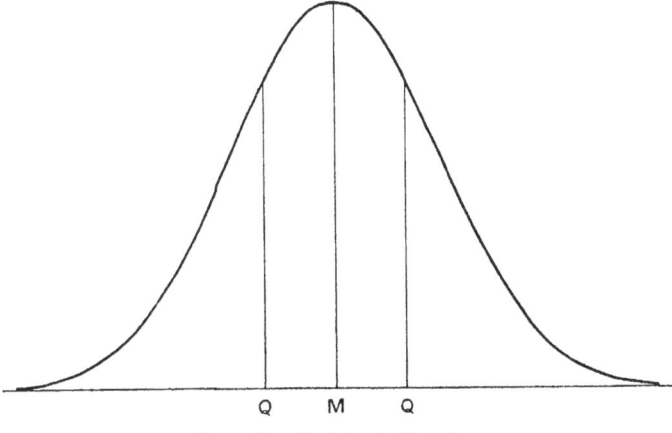

FIG. 8.—NORMAL CURVE.

base-line to meet the curve itself (M in the above figure). The curve is symmetrical on either side of the mode—that is to say, any two perpendiculars drawn from the base to the curve on either side of the mode and at the same distance from it will be equal in length.

When dealing with a symmetrical curve the position of the mode is identical with that of the *median*—the perpendicular line which divides the area of the curve into two equal halves, and the foot of this perpendicular also represents the *mean* or average of all the values from which the curve is constructed. In any actual case obtained by practical methods the position of the mode, the median, and the mean will only be approximately the same, because such a curve is never perfectly symmetrical.

The same curve can always be reconstructed if the position and magnitude of the mode are known, and, in addition, any one other point on the curve itself. A convenient point to take for this purpose is the point at which the curve is met by a straight line erected perpendicular to the base at such a distance from the median that it divides the area enclosed by the median, the base, and half the curve into two equal parts. The distance of such a perpendicular from the median is known as the *quartile*. Any given curve will have two quartiles one on either side of the median ; they are shown at Q and Q' in Fig. 8.

In practice an approximation to the normal curve of variability is constructed by plotting the values of a number of separate measurements or other determinations made upon different individuals. A *variate* is one of the separate numerical values from which a curve of variability can be constructed ; the biometrician usually deals with some such number as 1,000 variates. The total number of variates is represented by the area enclosed by the curve, and it will be seen that half the total number of variates falls between the two quartiles and half outside them.

A *class* (*cf*. p. 81) may be defined as a group of variates all of which show a particular value or a value falling between certain limits. The *frequency* of a class is the number of variates which it contains.

The amount of variation shown by a particular group of variates is measured by the degree of slope of the curve. A flat curve indicates greater variability and a steep curve denotes less variability. The flatter the curve—supposing the area (the number of variates) to remain the same—the further from the mode will be the position of the quartile, so that the distance of the quartile from the mode may be taken as a convenient measure of variability. In a theoretically perfect curve the distance of Q and Q' from M is equal. A curve obtained from an actual series of variates is never perfectly symmetrical, so that in practice the distance of Q and Q' from M may not be quite the same. In such a case the average of the two distances is taken as the measure of the variability of the material in question, and this value may be briefly denoted by the letter q.

In the example of variability of stature represented by Fig. 5, q is equal to 1·6 inches. This amount of variability can therefore be compared with other values representing the variability in stature and in other characters shown by various other groups of individuals. This, then, is the first important biometrical result which we have arrived at—the determination of a numerical value representing the amount of normal variability in any given case.

A measure of variability more often used than the quartile, especially in recent work, is what is known as the *standard deviation* of a normal curve, and may be expressed shortly as σ. σ represents a distance from the mode equal to $q \div 0\cdot6745$. Thus if σ is known, q can be readily determined, and *vice versa.* The reason for the more frequent use of σ is that it happens to be determinable with greater accuracy from an actual series of variates.*

The circumstance that half the total number of variates lies outside the limits of the quartiles and half within leads us to the consideration of what is known as the *probable error* The probable error of any statistical determination is a pair of values lying one above and one below the true value required—*e.g.*, the average stature of the whole of a race—such that it is an even chance that the value actually found will lie between them. Or the same thing may be expressed in another way. If we plot in the form of a curve a long series of actual determinations of a particular value, the probable error of a single determination will be nearly equal to the quartile of the curve so obtained. We may illustrate this state of things from our example of tossing coins, or still better by the essentially similar case of drawing balls out of a bag which contains a very large number of balls—black and white in equal numbers. Here the value to be determined experi-

---

* σ is found by multiplying the square of the deviation of each class from the mean (or mode) by the frequency of the class, adding together the series of products so obtained, dividing this number by the total number of variates, extracting the square root of the result, and multiplying by the number of units in the class range (this last number is very often unity). For further details with regard to the properties of the normal curve Davenport's ' Structural Methods ' may be consulted.

mentally is the relative number of black balls to white, which we know as a matter of fact to be equality ; and our single determination may consist in drawing out a hundred balls, which are afterwards returned to the bag. If we do this 1,000 times, and plot the number of black balls drawn each time, we shall arrive approximately at a curve having its mode at 50, and possessing a standard deviation which it is possible to determine from the instructions given in the footnote to p. 92. Multiplying σ by 0·6745 gives us the quartile, which represents the probable error of a single determination. That is to say, it is an even chance whether any single determination differs from 50 by more or less than q. In this particular example the quartiles would be found to lie very nearly at 46·6 and 53·4, so that the value of the probable error is 3·4.

The properties of the normal curve tell us a number of useful things about the probable error. In the first place its value varies inversely as the square root of the number of variates—that is to say, that in such a case as we have just described the probable error varies inversely as the square root of the number of balls drawn each time. We can realize this point more clearly when we remember that the linear dimensions of a curve vary with the square root of its area (the number of variates) ; the accuracy of our determination varies in fact with the quartile, which is the linear distance from the mode of a certain perpendicular.

We have seen that it is an even chance whether a single determination differs from the proper value by more or less than the amount of the probable error, an amount which we may denote by the letter e. The chance that any particular determination differs from the true value by more than twice the probable error is 4·5 to 1 against.

The chance that it differs by more than $3e$ is   21 : 1 against.
,,   ,,   ,,   ,, $4e$,,   142 : 1   ,,
,,   ,,   ,,   ,, $5e$,, 1,310 : 1   ,,

This is clearly very valuable information to possess when we are dealing with any kind of statistics.

We must now pass on to consider what methods are available to the biometrician for dealing with the problems of heredity. His way is to take a large number of pairs of relations, each pair consisting, say, of a father and a son, and to find out how much more like the members of such a pair are to one another on the average than the members of similar pairs of individuals would be, if taken at random and without regard to relationships from among the general population to which these fathers and sons belonged.

Now we shall see later on that this is not the only way of looking at the phenomenon of heredity, nor is it the way which is most familiar to biologists. But it is important to remember that what the biometrician means by amount of inheritance is a numerical value which expresses the average degree of likeness between a particular pair of relatives—for example, fathers and sons.

In the accompanying 'correlation table'—a purely imaginary illustration—there are tabulated the statures of 4,503 fathers, and those of one son of each of them. Thus 14 fathers, each 62 inches high,

TABLE IV.

HEIGHTS OF SONS (In inches) as rows; HEIGHTS OF FATHERS (In inches) as columns.

Heights of Sons	\[Fathers\] 62	63	64	65	66	67	68	69	70	71	72	73	74	75	76	Total Fathers for Sons of each Height	Modes of Arrays of Fathers
62	1	2	2	5	4											14	65·5
63	2	4	4	11	15	2	1		1	2						53	66
64	3	10	6	12	16	10	5	2	3	1						80	66·5
65	4	16	12	17	31	20	10	6	12	10	3	1				187	67
66	2	9	21	33	49	42	32	14	31	18	15	2	2			355	67·5
67	1	3	18	42	57	58	61	45	54	40	17	10	2			521	68
68	1	3	10	30	55	85	102	83	103	51	28	11	4	1		657	68·5
69		1	6	17	45	101	127	132	127	89	46	16	7	2	1	758	69
70		1	3	12	36	84	135	179	128	103	57	37	9	4	1	685	69·5
71		2	1	10	18	59	105	130	99	85	55	39	17	12	3	519	70
72			1	4	12	41	56	82	56	57	45	35	19	15	3	346	70·5
73				1	2	18	31	48	28	45	30	18	9	10	4	182	71
74					2	10	10	15	6	19	18	12	6	4	2	81	71·5
75						2	4	6	4	9	16	10	4	3	1	50	72
76							1	2	1	2	4	4	2	1	1	15	72·5
Total sons for fathers of each height	14	51	84	194	342	532	680	744	653	531	334	195	81	52	16	4,503	
Modes of arrays of sons	65·5	66	66·5	67	67·5	68	68·5	69	69·5	70	70·5	71	71·5	72	72·5		69

had 14 sons, whose heights are given in the first column. The series of heights of sons corresponding to a particular class of fathers is known as an *array*. Thus each column of the table represents an array of sons, and similarly each line represents an array of fathers.

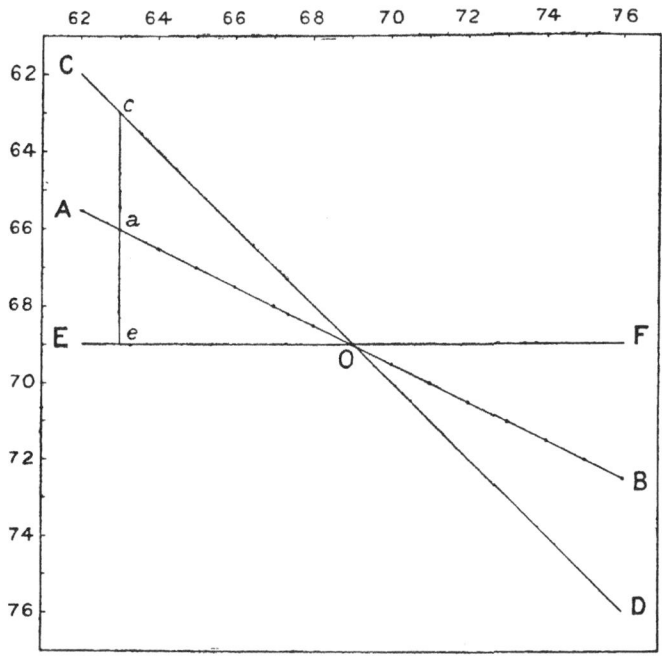

FIG. 9.—DIAGRAM OF CORRELATION.

The mode of each array of sons is given in the bottom line of the table.

Now if sons were on the average exactly the same height as their fathers, the modal value of each array of sons would be the same as the height of the corresponding class of fathers. If, on the other hand, there were no correlation between the heights of sons and those of their fathers the mode of every array of sons would be the same, and this value would be identical with the mode of the heights of all the sons taken at once. The actual result is found to be intermediate between these two possible extremes. Thus we see that sons tend to be like their fathers in respect of stature, but not exactly like, and if the example given were a real one the fundamental fact of a positive resemblance or correlation between the statures of fathers and sons would at once be clearly established.

The way in which a numerical value is attached to this correlation can be shown graphically.

In the diagram opposite, the dots indicate the values of the modes of the several arrays of sons as read off on the vertical scale to the left of the figure, the heights of the corresponding classes of fathers being read off on the horizontal scale. It will be seen that this series of dots lies nearly in a straight line which is inclined at a certain angle to the horizontal.

Now if there were perfect correlation between the heights of fathers and sons, or if the modal value of each array of sons were identical with the height of the corresponding class of fathers, the inclination of the line obtained in the above manner would be one of 45 degrees, as in the case of the line CD which passes through the points at which the values as read off in the vertical and horizontal scales are identical. If, on the other hand, there were no correlation the line would be horizontal, as EF.

The value taken to represent the amount of correla-

7

tion is the degree of slope of the line AB. This is expressed mathematically as tan *a*, *a* being the angle which the line in question makes with the horizontal.

When there is positive correlation this angle falls between 0 and 45 degrees, and tan *a* between 0 and 1. In the present instance tan *a* is 0·5. This value is known as the coefficient of correlation, and affords the basis of a numerical comparison with other similar coefficients obtained for other characters besides stature, and in the case of other pairs of relatives besides fathers and sons.

It ought now to be clearly understood that a complete resemblance between each class of fathers of a particular stature and the average stature of the corresponding array of sons would be indicated by the close approximation of our plotted points to a straight line making an angle of 45 degrees with the base line—a line, that is to say, having a slope of 1 in 1, or unity; whilst the entire absence of correlation would be represented by a line having no slope—that is to say, a horizontal line. The actual result in the example given is represented by a line having a slope of nearly 1 in 2, or 0·5.*

* *Correlated Variability.*—A precisely similar method is employed to measure the correlation of two parts or organs of the same individual. For example, the lengths of the right and left arms of men are very closely correlated. In order to attach a numerical value to this correlation the lengths of the right arms of a number of men are treated in the same way as the statures of fathers in the example given, and the lengths of their left arms in the same way as the statures of sons. The proper correlation coefficient can then be found by plotting the result; or the labour of plotting may be obviated by a process of calculation.

In the following table there are set down the correlation coefficients for stature in the case of seven pairs of relations, as obtained from actual data of a similar character to that already given by way of illustration.

TABLE V. (FROM PEARSON).

CORRELATION COEFFICIENTS FOR HUMAN STATURE.

Father and son ...	... 0·514
Father and daughter ...	... 0·510
Mother and son ...	... 0·494
Mother and daughter...	... 0·507
Brother and brother ...	... 0·511
Sister and sister ...	... 0·537
Brother and sister ...	... 0·553

Of the above, the first four values, representing correlation between parents and children, are seriously affected by the fact that the statistics from which they are derived show the existence of a marked correlation between husbands and wives in point of stature, amounting, indeed, to as much as 0·28—the result of what is technically described as selective mating. In the absence of such a relation between the statures of the parents, the correlation between parent and child might be expected to be distinctly less than that between pairs of brothers or sisters.

The term correlation replaces to some extent the older term *regression* employed by Galton. When speaking of regression the facts already described are regarded from a slightly different point of view. It is sometimes found convenient to speak of the regression of the mean stature of an array of sons toward the mean of the general population, instead of speaking of the correlation between the filial mean and the value of the parental class.

Regression represents the extent to which the average son is more like the mean of the general population than his father is. Correlation, on the other hand, indicates the amount by which the son is more like his parent than he is to the average of the general population. Thus, instead of being exactly like their parents, children are said to show regression towards the mean of the general population to which both parents and children belong.

In a case where the mean height of the fathers is identical with the mean height of the sons examined, and both are the same as the mean height of the general population, the coefficient of regression is simply equal to the reciprocal of the correlation coefficient between fathers and sons. In actual practice this condition is seldom realized, and it is then necessary to use a more elaborate method in order to determine the value of the regression coefficient.

Professor Pearson has extended the idea of correlation to the case of characters which are not capable of exact quantitative measurement. This extension is based upon the assumption that such characters follow a normal law of distribution in their variation, just in the same way as such a character as human stature was found to do. There is considerable doubt as to how far this assumption is justified, so that at the outset we may feel disposed to attach less importance to the actual values arrived at in this way than we should in

CORRELATION

normally. The method of calculation actually employed involves somewhat complicated mathematical
processes, but on Professor Pearson's authority we
may assume both the validity of the method and the
accuracy of the results obtained—so far as the actual
process of computation is concerned. For the purpose
of making the necessary calculations the data were
arranged in such a form as the following :

PARENTAL CORRELATION OF COAT COLOURS IN HORSES.

FILLIES.	SIRES.		Total.
Colour.	Bay and Darker.	Chestnut and Lighter.	
Bay and darker	631	125	756
Chestnut and lighter ...	147	147	294
Total	778	272	

By the suitable treatment of these figures the
value 0·45 was obtained as representing the coefficient
of correlation between sire and filly.

The amount of reliance which is to be placed in the
above method of determining the value of a correlation
coefficient was tested by arranging in a similar manner
data with regard to stature which had already been
treated in the form of a complete correlation table.
The whole number of fathers was divided into two
groups containing the individuals above and below a
certain stature, and the same was done in the case

of the sons. And the separation into two groups was made in several different ways by taking the dividing line between the groups at various heights. By applying to the statistics disposed in these various arrangements the same method as was applied to the statistics of horse colour already referred to, values varying between 0·52 and 0·6 were obtained for parental correlation; whereas the value arrived at by the more usual and reliable method was 0·514. It would therefore appear that there is with this method a tendency to obtain too high a figure, as compared with that derived from the method of the complete correlation table. When this source of inaccuracy is taken into consideration, in combination with the doubtfulness of the assumption upon which the method is based, it seems clear that its use will only give us a roughly approximate view of the correlation actually existing in the cases to which it is applied. Having made this reservation, we may compare the values given in the following table with those which appeared in Table V.:

TABLE VI. (FROM PEARSON).

AVERAGE PARENTAL CORRELATION.

Human eye colour	...	...	0·495
Horse, coat colour	...	...	0·522
Basset hound, coat colour	...	...	0·524
Greyhound, coat colour	...	...	0·507

AVERAGE FRATERNAL CORRELATION.

Human eye colour	...	...	0·475
Horse, coat colour	...	...	0·633
Basset hound, coat colour	...	...	0·524
Greyhound, red in coat	...	...	0·700
Greyhound, black in coat	...	...	0·660

Thus if we use the term inheritance at present simply to express the fact that a more or less definite numerical value can be attached to the average amount of resemblance between any specified pair of relatives, we see that a considerable number of physical characters appear to be inherited at approximately the same rate in men and in animals.

More than this, Professor Pearson has shown, by the use of the same method as was applied to the case of physical characters not quantitatively measurable, that the average resemblance in *mental* characteristics between pairs of brothers, pairs of sisters, and pairs made up of a brother and a sister, can be expressed by the values given in the following table :

TABLE VII. (FROM PEARSON).

Character.	Brothers.	Sisters.	Brother and Sister.
Vivacity	0·47	0·43	0·49
Assertiveness ...	0·53	0·44	0·52
Introspection ...	0·59	0·47	0·63
Popularity	0·50	0·47	0·49
Conscientiousness ...	0·59	0·64	0·63
Temper	0·51	0·49	0·51
Ability	0·46	0·47	0·44
Handwriting... ...	0·53	0·56	0·48
Mean	0·52	0·51	0·52

A sample of the collected facts from which this information is arrived at is given in the table on p. 104.

CONSCIENTIOUSNESS : BROTHER—BROTHER.

| SECOND BROTHER. | FIRST BROTHER. | | Total. |
	Keen.	Dull.	
Keen	970	216·5	1,186·5
Dull	216·5	287	503·5
Total	1,186·5	503·5	1,690

Every child was classified in this way as being either above or below an average standard in respect of each character. The estimations were made by teachers having at least six months' experience of the children in question.

The method of statistical treatment was, as we have said, the same as that employed in the case of physical characters not capable of quantitative measurement, and there is little doubt that it is equally valid in the present case. We may well feel, however, some hesitation in accepting as sound the data to which the method is applied. At the best this data can only be of a roughly approximate kind. The evidence is, however, undoubtedly sufficient to establish the conclusion that mental characters are inherited in man, and that they are probably inherited at a rate not greatly different from that at which physical characters are inherited. For it will be observed that the values given in Table VII. are in close agreement with one another, and that they also agree with the average value of fraternal correlation as found for a variety of physical characters both in men and in other animals. Assuming—and the assumption seems to be a reasonable one—that equal fraternal correlations indicate the existence of equal correlations between parents and children, we arrive at the conclusion that the resemblance between parents and their offspring is of much the same kind and amount in the case of mental as it is in the case of bodily characteristics.

What we may perhaps describe as the main generalization so far arrived at by biometricians is known as the Law of Ancestral Heredity. This hypothesis supposes, or at least in its original form supposed, that every ancestor of a particular individual contributes its quota to the heritable qualities displayed by that individual. The law also states that the average amount of resemblance between an individual and any particular ancestor is capable of definite numerical expression. Thus the mean amount of correlation between (1) the two parents and the offspring, (2) the four grandparents and the offspring, (3) the eight great-grandparents and the offspring, and so on, is believed to diminish in a geometrical series, which is the same for all organisms and for all characters. The actual amounts of these correlations were expressed by Galton in the form of the series 0·50, 0·25, 0·125, etc. Pearson regards them as being more nearly represented by the more rapidly diminishing series 0·6244, 0·1988, 0·0630, etc.

Now, there can be no doubt that the law as stated above has been disproved in specific instances, and was indeed disproved by the work of Gregor Mendel before ever it was enunciated, although Mendel's work was not generally known until later. According to Mendel's theory of inheritance, certain ancestors contribute *nothing* to the constitution of certain offspring in respect of certain characters. Furthermore, the modification of the law of ancestral heredity which applied to alternative inheritance, and which was assumed in working out the inheritance of coat colour in thoroughbred horses, has recently been shown not to apply to that particular case.

Unfortunately, most of the further biometrical generalizations are based upon the assumption that the law of ancestral heredity is strictly true. So that whilst we have spent some time in considering the facts of normal variability and of correlation between relatives, because these facts are quite independent of any theoretical assumption, the remainder of our review must be passed over at a more rapid rate. Until the theoretical conclusions now to be described have been revised by their authors in the light of recent knowledge, it is difficult to say how much reliance is to be laid upon them, but it seems quite likely that they will hold good as approximations. Indeed, though not applying to individual cases, the law of ancestral heredity does seem to hold good as a statistical statement of general results, so that there would be no objection to it on either theoretical or practical grounds if only it had been enunciated in some such terms as 'a law of average ancestral resemblance.' Thus it is quite possible that the total contribution of the eight great-grandparents of an individual may be on the average correctly represented by Pearson's fraction, even though their individual contributions are not always the same.

Let us, then, briefly examine some of the further conclusions which have been drawn from the data of the biometricians.

Assuming the law of ancestral heredity, Pearson has arrived at very interesting conclusions with regard to the effects of artificial selection when the correlation coefficients have those values which have been actually found for them in the case of the human race.   In the statement which follows, ancestors are supposed to have been selected showing in each generation a deviation h from the general mean of the population.   Thus, suppose the character selected to be stature : suppose the mean height of the population to be 6 feet, and the selected individuals to be 6 feet 6 inches high ; h is then 6 inches, and only individuals of a height of 6 feet 6 inches would be selected as parents in each generation, so that after three generations of selection we should be dealing with children whose parents, grandparents, and great-grandparents were all of this particular height

Pearson calculates that after one generation of selection the immediate offspring will show 0·62 of the character selected (0·62 h).   After two generations they will show 0·82 h, after three 0·89 h, and after a great number of generations 0·92 h.   Thus in a comparatively small number of generations the development of a character may be raised to within 90 per cent. of the value selected, but, after this, further selection has very little effect. If selection is stopped after one generation, and the selected stock is then inbred, it was calculated that the first generation of inbred stock would show 0·59 h, the second 0·56 h, the third 0·52 h, and the tenth 0·35 h. If, on the other hand, inbreeding was started after the selection had continued for a large number of generations, the first generation of inbred stock will show 0·86 h, the second 0·81 h, the third 0·77 h, and the tenth 0·51 h. So that inbreeding of a selected stock is followed by a very gradual return towards the mean character of the original race.*

It must be remembered that in the calculation which led to this result perfect normal variability was assumed, and the contribution of every ancestor of the same degree to the hereditary endowment of the offspring was supposed to be exactly equal. Since both these assumptions are very unlikely to be realized in any actual case, the statement here given must only be regarded as an approximate indication of what is likely to take place.

Some remarkable observations have recently been published by Professor Johannsen, of Copenhagen, and from them are drawn conclusions which seem likely to lead to a distinct advance in our understanding of the process of so-called continuous variation, and of the way in which such variations are transmitted.

* From this it seems necessarily to follow that it is impossible to establish a permanent breed simply by a process of selection.

Johannsen's experiments have so far only been published in the form of a preliminary abstract, but his conclusions are of such interest that it seems necessary to draw attention to them. The proviso must be made, however, that further evidence is necessary in order to justify their complete acceptance.

The experiments in question were made upon plants which could be self-fertilized for a series of generations. In this way many complications were avoided which are inevitably introduced in the case of biparental inheritance. Barley and kidney beans were among the plants examined, and the simplest character considered was the size of the seeds of the latter as measured by weighing. In this particular series of experiments each plant was regarded as being characterized by the average weight of the seeds which it produced.

All the descendants arising from a single plant by self-fertilization are spoken of by Johannsen as making up a ' pure line.' And the members of such a line showed, in respect of the weight of their seeds, normal variability about a mean or type value. The general population of bean plants, made up of a great number of such pure lines, also exhibited a normal curve when the weights of the seeds were plotted. The pure lines composing such a population showed various types, some of them close to the modal value of the population, but others differing widely from it. If now a somewhat widely deviating member of a particular line was selected for propagation, its offspring showed regression to the type of this particular line, and not to the mean value of the general population.

The case is indeed precisely similar to the supposed example of a mixture of races of peas, which was made use of as an illustration at the beginning of the present chapter. In other words, a pure line consists of a group of individuals which has a normal variation of its own, and the offspring of which by self-fertilization breed true to the type of their own particular group, and show no regression towards the type of the general population to which the group belongs.

If we were to carry on this conception to the case of bisexual inheritance, we should find that the different pure lines would become crossed and confused together in a way which would be very difficult to disentangle. There is no reason to doubt that statistical treatment of such a population would yield similar results to those actually obtained by biometricians from the data at their disposal; and we may notice that a fortuitous mixture of a considerable number of pure lines, having slightly different types, would admirably ulfil the conditions we have seen to be necessary in the case of material, to which methods based upon the theory of chance are to be applied. The phenomena which follow upon the crossing together of two or more pure lines still remain to be worked out, but it is not unlikely that they will be found to conform to those laws of heredity associated with the name of Mendel which are explained in Chapter VII. If this should be found to be the case, it is not impossible that the theory of pure lines, in combination with the method of inheritance referred to, may adequately serve to describe those phenomena to account for which the law of ancestral inheritance was called into existence.

The conclusions to which Professor Johannsen's experiments lead him may be summed up as follows: Individuals which differ (in size, for example) from the mean of a population give rise to offspring which differ from that mean value in the same direction but to a smaller extent. Selection, therefore, will produce a change in the average character of a population taken as a whole. Selection *within a pure line* produces no effect of this kind. The average character of the offspring of typical members of the line is the same as that of the offspring of members which show the widest deviations from the type.

Selection *in a population* consists in the partial separation of those lines the types of which differ in the required direction from the average character of the population. This effect must of necessity come to an end when the most eccentric line is completely isolated. The great complications introduced when the lines are intermingled through mixed breeding may make this process of isolation a very tedious one.

It will be seen that the values calculated by Pearson to represent the result of selection in a population agree quite well with Johannsen's explanation of the constitution of such a population out of a number of pure lines. The result of Professor Johannsen's further experiments will therefore be awaited with great interest by biologists and biometricians alike.

On the theory of pure lines it is to be noticed that the personal character of a particular ancestor has no influence upon his descendants ; it is only the type of the line to which he belongs which influences the offspring, so that this theory is in perfect agreement with Weismann's theory of inheritance as described on p. 68.

It is also to be observed that the principle of the pure line applies only to quantitative characters—such characters of size, or of weight, or of proportion, as are very seldom made use of by systematists for the distinction of natural species.

Hugo de Vries.

[To face p. 113.

CHAPTER V

MUTATION is the term applied by de Vries to express the process of origination of a new species, or of a new specific character, when this takes place by the discontinuous method at a single step—a process which he regards as the most important if not the sole method by which new species or specific characters arise. We shall see that although de Vries has recently done much to forward the propagation of this idea, the belief that such a discontinuous process is the normal method by which new species come into existence has been developing for a considerable time.

We have seen that those who accept the idea of evolution by the action of natural selection upon a series of minute and almost imperceptible variations are confronted with the difficulty of explaining how by this method there could arise a number of different structures or parts so co-ordinated as to share in a common function. Moreover, a closer examination of the actual processes of variation and inheritance render it doubtful whether the selection of continuous variations of even a simple character can ever lead to the development of a permanent new race. The result of Pearson's calculations, described in the preceding chapter, seems to indicate that the selection of a certain value of a particular character for many generations will never lead to the formation of a race in which the mean value of the character is as high as the selected value. But, says the selectionist, it will happen in Nature that as the standard of the race is raised by selection, the value selected will be still further raised, and so on, and in this way an indefinite amount of improvement is rendered possible. If Johannsen's conclusions are well founded, this is clearly not the case ; on the contrary, there is a perfectly definite limit to the effect which selection can produce.

The question whether or not a gradual method of evolution is possible has not yet been absolutely decided for any single species or character, but it certainly seems that now for the first time the possibility of a definite decision is within sight. At the same time it is impossible to prove a universal negative If we look at the other side of the problem we shall find that the evidence in favour of an alternative process has multiplied even faster than the evidence against the continuous accumulation of minute differences ; and the present tendency is certainly to look for other sources of specific distinctness than that which is offered by the natural selection of continuous variations.

Even before the new evidence which we have briefly outlined was available, Herbert Spencer found the difficulties in the way of accepting the purely

Darwinian explanation to be so great, that he adopted the hypothesis of the inheritance of acquired characters, as being the only adequate explanation of the phenomena which was in his time available.

Unfortunately, satisfactory evidence that such a form of inheritance ever actually takes place has never been forthcoming in sufficient amount to lead to universal conviction. Indeed, at the present day the consensus of opinion among experts is undoubtedly to the effect that acquired characters are not inherited at all, except in so far as better nutrition of the parent may lead to the production of more vigorous off-spring. And it seems clear that such an effect as the latter cannot go on accumulating for more than a few generations.

Thus we see that in the purely Darwinian view there is something wanting, whilst the Lamarkian explanation is ruled out of court for the present for lack of evidence. If, at this point, we find that in Nature a co-ordinated set of structures can and does arise in an already perfected condition at a single step, and that such phenomena take place with sufficient frequency to give ample opportunities for the survival of the new type so arising, we have at once discovered an alternative way out of the difficulty. Such a discovery must throw abundant light on the obscurity overshadowing the methods by which evolution has taken place, even though we may not yet have arrived at any kind of explanation of the cause of this phenomenon of co-ordinated and definite variability.

The actual observation of variations of this kind is

8—2

of quite recent date, and their recognition is largely
due to the exertions of Bateson. But the idea that
this is the way in which evolution takes place is very
ancient, as a few quotations will clearly demonstrate.

The idea that definite structures may arise, each as a
whole and in a perfect condition, was clearly propounded
by Aristotle in a passage which it is a little curious to
find quoted at the beginning of the 'Origin of Species.'
Darwin's note is to the following effect : After remarking that rain does not fall in order to make the
corn grow, any more than it falls to spoil the farmer's
corn when threshed out of doors, Aristotle adds, ' So
what hinders the different parts of the body from
having this merely accidental relation in nature ? as
the teeth, for example, grow by necessity the front
ones sharp, adapted for dividing, and the grinders
flat, and serviceable for masticating the food ; since
they were not made for the sake of this, but it was the
result of accident. And in like manner as to the other
parts in which there appears to exist an adaptation
to an end. Wheresoever, therefore, all things together
—that is, all the parts of one whole—happened like
as if they were made for the sake of something, these
were preserved, having been appropriately constituted
by an internal spontaneity ; and whatsoever things
were not thus constituted, perished and still perish.'

Upon the above passage from Aristotle Darwin
comments as follows : ' We here see the principle of
natural selection shadowed forth, but how little
Aristotle fully comprehended the principle is shown
by his remarks on the formation of the teeth.' We

may state at once that these very remarks upon the formation of the teeth almost exactly embody the views of the modern mutationist.

Perhaps the earliest use of the actual word 'mutation' in this sense is to be found in 'Pseudodoxia Epidemica,' by Dr. Thomas Browne. I quote from Chapter X., 'Of the Blackness of Negroes '* (second edition, 1650): 'We may say that men became black in the same manner that some Foxes, Squirrels, Lions, first turned of this complection, whereof there are a constant sort in diverse countries ; that some chaughes came to have red legges and bills, that Crows became pyed ; All which *mutations*, however they began, depend upon durable foundations, and Such as may continue for ever.'

The experiments upon cross-breeding, which are described in a later chapter, will be found fully to bear out the idea that 'mutations,' or definite characteristics which have arisen in a definite way, do depend upon durable foundations.

The late Professor Huxley's emphatic approval of the 'Origin of Species,' as signalized in his reviews of the first edition of that work, was tempered by the following mild criticism : 'Mr. Darwin's position might, we think, have been even stronger than it is if he had not embarrassed himself with the aphorism "Natura non facit saltum," which turns up so often in his pages. We believe . . . that Nature does make jumps now and then, and a recognition of the fact is

* I am indebted to my friend Mr. R. C. Punnett for this reference.

of no small importance in disposing of many minor objections to the doctrine of transmutation.'*

The first person to formulate a more or less precise view upon the subject of definite variation was Francis Galton, although this author never entered into the question at any great length. Galton's attitude towards the problem in its early stages may be gathered from the following quotation from his ' Natural Inheritance ': ' The theory of natural selection might dispense with a restriction for which it is difficult to see either the need or the justification—namely, that the course of evolution always proceeds by steps that are severally minute, and that become effective only through accumulation. That the steps may be small, and that they must be small, are very different views ; it is only to the latter that I object, and only when the indefinite word " small " is used in the sense of " barely discernible," or as small compared with such large sports as are known to have been the origins of new races.'†

But more than this, the idea of the existence of stable forms, such as may be supposed to have arisen by large and sudden variations, is very well expressed by Galton in his division of varieties into the three groups of primary types, subordinate types, and mere deviations from the latter. A most luminous analogy is afforded by the three types of public vehicles which at the end of the nineteenth century were characteristic of the streets of London ; and it is impossible to resist quoting Galton's account of them. These three

* ' Collected Essays,' vol. ii., p. 77.
† ' Natural Inheritance,' p. 32.

kinds of carriages, 'namely, omnibuses, hansoms, and four-wheelers, are specific and excellent illustrations of what I wish to express by mechanical types as distinguished from subtypes. Attempted improvements in each of them are yearly seen, but none have as yet superseded the old familiar patterns, which cannot, as it thus far appears, be changed with advantage, taking the circumstances of London as they are. Yet there have been numerous subsidiary and patented contrivances, each a distinct step in the improvement of one or other of the three primary

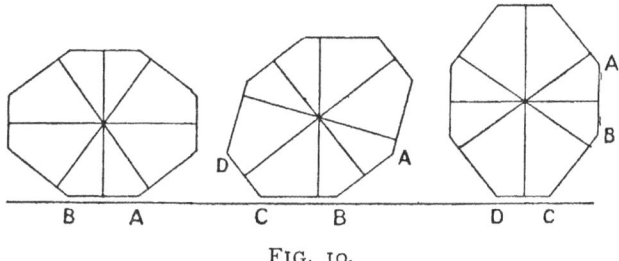

FIG. 10.

types, and there are or may be an indefinite number of varieties in details, too unimportant to be subjects of patent rights.'*

More recently Galton might have pointed out the introduction of motor traffic as illustrating a distinct mutation.

The distinction between primary and subordinate positions of stability is further excellently illustrated by the model which is here represented, and which is known as Galton's polygon (Fig. 10).

* 'Natural Inheritance,' p. 26.

The first position of the model, resting upon the side A B, may be taken to represent the condition of a type or stable form. A comparatively small push (variation) will lead to the production of the subtype illustrated by the position B C. When in this new position, it is easier to cause the model to return to its original position A B than it is to make it pass on to the new and more modified position resting upon the side C D. A strong push (mutation) may force the model to pass through the position C D until it comes to rest on the side opposite to A B. This fresh position represents a new stable form, and it is now once more surrounded by positions of subordinate stability—subtypes.

One more analogy before we pass on to consider the more recent observations upon discontinuous variations or mutations. We may compare the difference which exists between deviations and stable forms, arising by fluctuating and by definite variation respectively, with the behaviour of the atoms of chemistry, as expressed in the account of their structure recently given by Professor J. J. Thomson. Such an atom is regarded as being made up of a number of electrons or corpuscles bearing definite relations to one another in space. In certain circumstances it seems that it may be possible to remove a series of these corpuscles from the atom one at a time, in which case every such successive removal would be accompanied by a comparatively gradual and progressive change in the properties of the atom so modified. But after a certain time a point would be reached at which the removal of one more electron would necessitate a complete rearrangement of the remaining corpuscles in order to arrive at a new position of equilibrium, and this change would be accompanied by a marked alteration in the chemical properties of the atom itself. In like manner the chemical composition of the living substance of a race of organisms may be conceived to alter step by step, every such step being accompanied by comparatively unimportant changes in its visible characters, until the time arrives when any further alteration must be associated with a deep-seated revolution in the constitution of the living substance, and with a corresponding marked mutation in the external features of the members of the race.

The first really definite attempt to collect and co-ordinate the facts of discontinuous variation was made by Bateson in his book entitled 'Materials for the Study of Variation,' published in 1894. The introduction and concluding remarks at least of this volume ought to be read by everyone who is interested in these subjects. The bulk of the book contains a mass of material of great value to specialists.

After pointing out the difficulties which prevent his acceptance of the orthodox belief in the origin of discontinuous and apparently adaptative types of animals and plants through the action of natural selection on minute variations, difficulties to which we have already paid some attention, Bateson records his conviction that the facts of discontinuous variation afford a way out of the difficulty. He shows (1) that differences of the kind which are generally used to distinguish separate species may arise as single variations ;

(2) that such a form of variation is by no means so uncommon a phenomenon as was formerly supposed ; and (3) that variations of this kind may occur in every description of organ and part in a number of different plants and animals. The facts with which the main bulk of the book is concerned have reference to the animal kingdom.

We shall find it profitable to consider the views expressed in this book a little more closely, though it would occupy too much space to give even a brief summary of the facts upon which they are based, and for which reference must be made to the original.

In the first place Bateson calls attention to the phenomenon of *symmetry* as being a characteristic feature common to almost all organisms. This symmetry may manifest itself in a number of different ways. In *bilateral* and *radial* symmetry the parts symmetrically disposed are related to one another in the same kind of way as are an object and its image reflected in a plane mirror. Such symmetry, as, indeed, every kind of symmetry, is usually associated with a repetition of parts. In the present instances the parts are repeated in pairs, as with the two eyes in the human face ; or in a radial series, like the arms of a star-fish, or the petals of a buttercup. To this phenomenon of the repetition of parts, generally occurring in such a way as to produce a symmetry or pattern, the term *merism* is applied.

Symmetry may affect the proportions and shape of the body of an animal or plant as a whole, or, on the other hand, separate parts or organs may show a separate symmetry of their own. For the phenomena thus distinguishable separate terms are proposed. A *major symmetry* is a form of pattern which includes the body as a whole, as in the case of most animals where the two sides of the body closely resemble one another. A *minor symmetry* is a pattern completed in a separate organ or part—for instance, in the flower of a plant or the limb of an animal.

Once more we may lay stress upon the universal existence of pattern among living things. Bateson points out that in collecting any kind of living creature it is the symmetry of it which, as a general rule, first catches the eye and distinguishes the organized body from surrounding inanimate objects.

The phenomenon of merism or repetition of parts being understood, we are in a position to consider the subdivision of variations into *meristic variations* and *substantive variations* respectively.

Meristic variations are variations in symmetry and in the number of repeated parts. A change in the number of organs in a series may conceivably take place gradually by the addition or subtraction of successive fractions of a part. But, as a matter of fact, this is very seldom the case. The increase or decrease usually involves one whole member at a time and sometimes more, so that this kind of variation is, as a rule, discontinuous. Abundant illustrations of this fact are to be found in the case of changes in the number of such parts as the teeth or vertebræ of mammals ; and a particularly good instance is afforded by the variations which take place in the number of ray florets in various composite plants—*e.g.*, the daisy and chrysanthemum. It is suggested that meristic variations are connected with definite changes in the mechanical relations of dividing parts, and that it is in the mechanics of cell-division that the explanation of their discontinuous appearance is to be sought for.

Thus when, for example, a tulip-flower appears having its parts perfectly developed in sets of four instead of in sets of three, it is suggested that the arrangement in fours, like the arrangement in threes, fulfils certain conditions of equilibrium among the forces which affect the cell-divisions in the rudiment of the flower, and that these conditions of stability would not be equally well provided for by any intermediate arrangement.

Substantive variations are changes in the actual constitution or substance of the parts themselves. For example, a plant with coloured flowers may give rise to offspring the flowers of which are white. There seems to be no mechanical necessity for such variations to be discontinuous rather than continuous ; it is quite possible to imagine a gradual dilution of colour taking place throughout a long series of generations. Discontinuous substantive variations are, however, not infrequent, and in such cases it is suggested that they may be associated with definite changes in chemical composition. Thus, for example, definite alterations in the colour of offspring as compared with their parents seem almost necessarily to be of this nature.

The further evidence contained in the book we are considering refers entirely to meristic variation.

An important point with regard to repeated parts is to be observed in the fact that in a pair of allied species, in which a series of repeated organs in the one is clearly comparable with a similar series in the other, all the parts in one form may differ from those in the second by the same kind of distinction, whether this be qualitative or numerical. The facts suggest strongly that such cases are to be accounted for by all the parts in question in one or both species having varied in a similar way at the same time rather than in succession. The occurrence of such a similar and simultaneous process of variation of repeated parts clearly simplifies in a marked degree the process of evolution, and greatly reduces the time which would be required for this process, if similar changes in repeated parts always took place successively. If we take an extreme case the latter supposition becomes absurd. In the albino or pure white types which occur as variations in many species of birds and mammals it is obvious that every hair or feather has taken on the white colour at the same time and for the same reason, whatever that reason may have been. Hairs or feathers are very good examples of repeated parts of the kind of which we have been speaking. It appears, too, that colour patterns may originate and change in a similar manner. In the case of such a bird as the peacock we should expect on this view that the pattern varied in all the tail feathers simultaneously, nor is it necessary to suppose that even this process took place by a very long series of minute steps. If we find that the splendid coloration of the peacock's tail arose by a few marked variations, each of which occurred simultaneously in all the feathers at once, several serious difficulties are avoided, and on the analogy of similar known cases we have every reason to believe that this was so. And similar changes may take place in cases where the pattern depends on the coloration of a group of feathers or hairs. Indeed, if we consider, we shall find it very difficult to picture such a process as taking place in any other way. We can scarcely suppose the spots of the leopard, for instance, to have arisen one at a time.

An important kind of discontinuous variation is that to which Bateson has applied the term *homœosis*. The same sort of change had previously been described by Masters in the case of plants under the name 'metamorphy,' but the latter expression has also been employed in other senses. Homœosis consists in the assumption by one member of a meristic series of the form or character proper to another member of the same series ; for example, the modification of the petal of a flower into a stamen, or of the eye of a crab into an antenna-like organ.

' In these cases a limb, a floral segment, or some other member, though itself a group of miscellaneous tissues, may suddenly appear in the likeness of some other member of the series, assuming at one step the condition to which the member copied attained presumably by a long course of evolution.' *

The phenomenon of homœosis is frequently to be seen among the parts of flowers. Double flowers in

* 'Materials for the Study of Variation,' p. 570.

many cases—for instance, in the case of the rose—arise by the development of petal-like organs in the position which would properly be occupied by stamens. A parallel process is to be seen in the heads of composite flowers, such as the chrysanthemum. In a double chrysanthemum the florets of the disc develop in the likeness of ray florets. Both these cases would be classed as examples of *outward homœosis*, because the parts concerned resemble organs normally developed in a whorl exterior to themselves. A case of *inward homœosis*, on the other hand, is afforded by the appearance of a petaloid calyx—for example, in a tobacco-plant—the outermost whorl of the flower taking on the appearance of a whorl internal to itself.

In cases such as these we observe once more the occurrence of a marked and definite change, which, though at first sight quite distinct from the method of similar and simultaneous variation, yet bears a certain resemblance to that process in the fact that the direction in which a particular part varies is not wholly unrelated to the behaviour of other parts of the same organism. The process thus briefly described seems likely to have had considerable importance in evolution, notably in the origin of differences in the numerical relations of the bones in various parts of the spinal column in different vertebrate animals.

The preceding account of the conclusions drawn from Bateson's laborious study of variation has involved a good deal of technicality, but this is, unfortunately, unavoidable. The point chiefly to be emphasized is the frequent occurrence in Nature of

variations of a definite or discontinuous type—the
fact that differences of the kind which are constantly
used to distinguish natural species can and do arise
in Nature at a single step, so that it is not necessary
for such differences to be built up gradually by the
action of natural selection.

De Vries, in his ‘ Mutations Theorie,’ goes further
than this, and attacks the position held by those who
accept the doctrine that natural selection of individual
differences can ever lead to definite and permanent, or
specific, distinctions.   Indeed, one of the chief contributions of this author to the species controversy is to
point out that the belief that artificial selection acts
in this way upon domestic plants is based upon a mis-
apprehension.   De Vries himself has carried out a
number of experiments in selection, and he comes to
the conclusion that selection of ordinary individual
differences has no permanent effect at all.*   The actual
effect of this kind of selection is well illustrated by the
results of the processes employed in the sugar-beet
industry, in which elaborate care is taken to select
those roots which contain the highest percentage of
sugar for the purpose of propagation.   This process
was followed at first by a rapid improvement, but the
rate at which the percentage of sugar increased soon
fell off, until at the present day all that selection can
effect is to keep up the standard of excellence already
attained.   Moreover, that this process of improvement

* Compare, however, Johanssen's more recent conclusions
see p. 111).

was a very gradual one is to be accounted for in part, at least, from the fact that the methods of selection themselves gradually improved from year to year. There is no reason to doubt that a thoroughly efficient method of selection would have worked its full effect in a few generations. A similar state of things is said to be the case with the cereals, such as wheat and barley, which have been selected largely for the size of the grains. From his own experiments, de Vries has come to the conclusion that, when selection is really efficient, the full possible effect of this process is exhausted in quite a small number of generations, and that then the only further effect of selection is to keep up the standard already arrived at.

We have seen that the theoretical conclusions of the biometricians are in agreement with the opinions here expressed, so long as selection is understood to be confined to the choosing out of parents which show a definite standard value of the character under consideration, this value being the same in each generation. Under these circumstances, Professor Pearson concludes that in the first two or three generations a marked advance in the desired direction will take place, but that further selection (in this sense) will have comparatively little effect. But the believer in continuous evolution maintains in addition that selection will be followed to an indefinite extent by further variations in the direction of selection, since otherwise selection could never lead to important changes in organization. In the face of the strong contrary evidence, and of the fact that alternative

9

methods of evolution are now known to be available, the burden of proof of this proposition seems to lie with those who maintain the all-important influence of continuous variation and selection. At present we are free to reply in the words of Malthus, who long ago protested against the extravagant powers which were ascribed to the selection of small differences.

'I have been told,' Malthus writes, 'that it is a maxim among some of the improvers of cattle that you may breed to any degree of nicety you please, and they found this maxim upon another, which is, that some of the offspring will possess the desirable qualities of the parents in a greater degree. In the famous Leicestershire breed of sheep, the object is to procure them with small heads and small legs. Proceeding upon these breeding maxims, it is evident that we might go on until the heads and legs were evanescent quantities ; but this is so palpable an absurdity that we may be quite sure the premises are not just, and that there really is a limit, though we cannot see it or say exactly where it is.' *

The only recorded example I am aware of in the case of animals, which shows the result of long-continued selection acting upon a quantitative character, is afforded by the case of the American trotting-horse. In this case it appears highly probable that we are dealing with a character which varies in a strictly continuous fashion. In his book upon ' The Trotting and Pacing Horse in America,' Hamilton Busbey gives a table from which the diagram on the opposite page

* 'Essay on Population,' 6th ed., vol. ii., p. 11.

is constructed. The entries in this table show the
fastest times recorded for the feat of trotting a measured
mile in various years beginning with 1818. The vertical scale contains the times, which vary from three
minutes down to one minute fifty-six seconds, and the
horizontal scale shows the year in which the record was

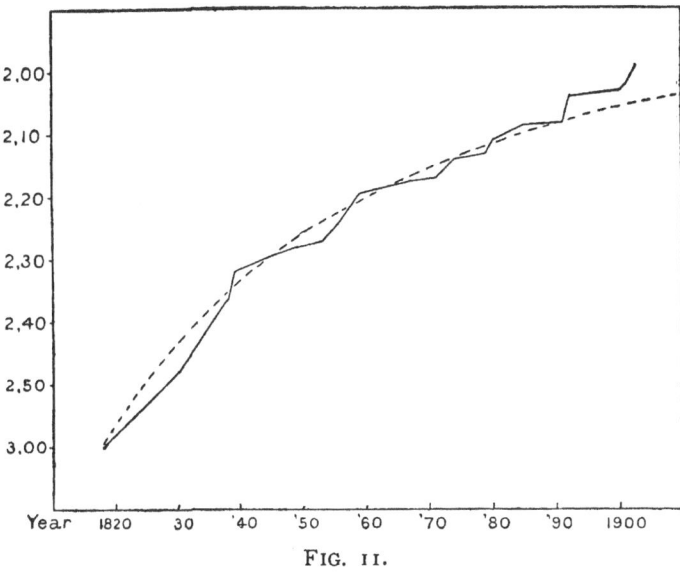

FIG. 11.

The figures to the left of the diagram are to be read as minutes and
seconds.

made. Some part of the improvement shown is clearly
to be associated with better tracks, improved methods
of training, etc., but these will scarcely affect the
general character of the improvement due to selection. As may readily be seen from the diagram, the
improvement is at first rapid, but afterwards becomes

gradually slower and slower. At the end of the series two sudden steps upward break the general regularity of the series of records. But on examination of the evidence it is found that these are associated with special conditions, and are not really exceptional. The first of these breaks—that which occurs in 1892—is coincident with the introduction of a new type of sulky, having ball bearings and other improvements ; whilst the record of 1903 was accomplished behind a pacemaker carrying a wind-shield. Neither of these records, therefore, is strictly comparable with the rest of the series.

The observations in this case do not, indeed, seem to be sufficient to afford the basis for a final decision against the theory of the indeterminate power of selection. Yet Malthus' criticism clearly applies very definitely to such a case—*i.e.*, there must be a limit beyond which the speed of the trotting-horse will never improve without a fundamental change taking place in his organization. It seems, therefore, safe to conclude that the curve to which the series of recrods approaches is of the character of a parabola—*i.e.*, one which continually becomes more and more nearly horizontal as the speed of the horse gradually approaches its highest possible limit.

De Vries, then, contends that all new domestic breeds have arisen by the discontinuous method as definite novelties. Darwin himself was perfectly aware that this is usually the case, but the conclusion which he drew from the fact was a different one, as the following passage shows :

' He (man) often begins his selection by some half-monstrous form, or at least by some modification prominent enough to catch the eye or to be plainly useful to him.' But he goes on : ' Under Nature, the slightest differences of structure or constitution may well turn the nicely-balanced scale in the struggle for life, and so be preserved.' *

Of the origin of a new type of plant in this definite and sudden fashion, the Shirley poppies afford an excellent example. These originated in a mutation of the common wild field-poppy (*Papaver rhœas*). In 1880 the Rev. W. Wilks, Vicar of Shirley, near Croydon, noticed among a patch of this plant growing in a waste corner of his garden a solitary flower, the petals of which showed a very narrow border of white. The seeds which this flower produced were sown, and next year, out of about two hundred plants, there were four or five upon which all the flowers showed the same modification. From these, by further horticultural processes, the strain of Shirley poppies originated. We may point out in passing that if the original plant had been self-pollinated, a much larger proportion of the new type might have been expected to appear in the next generation.

In the course of his own experiment, de Vries has obtained quite a number of new types of plants by methods like the above. It is to be observed that the novelty in these cases usually shows a considerable range of normal variability of its own, and that its first appearance is generally in the form of an extreme

* ' Origin of Species,' 6th ed., p. 60.

negative variation* from its own proper type.   In this
way the novelty may not appear to be very far removed
from a high normal variation of the original type.   The
behaviour of the progeny of the two types, however—
types which might thus in themselves be readily confused—is entirely different, and a ready means of distinguishing them is thereby provided.   Each set of offspring shows regression to its own proper modal value ;
so that the offspring of the novelty show a further
marked development of the new features, whilst the
offspring of an extreme normal variation resemble the
type of the original form more closely than they do
their own immediate progenitor.

If new types are not produced among domesticated
productions by the action of artificial selection, and all
that selection can effect is to pick out definite novelties
when they occur, the analogy between natural selection and artificial selection breaks down, and a large
and important section of the evidence in favour of the
production of natural species by the action of natural
selection is destroyed.   In the place of this explanation de Vries would put the theory of mutation, according to which new species arise by single steps as
definite novelties, just in the same way as we find that
domestic varieties are produced.   More than this, de
Vries believes that he has discovered a set of new
species in the very act of originating from an old
one in this way, a discovery which affords the basis

* *I.e.*, a variant belonging to a class situated some distance from the mode of normal variability of the novelty,
and on the side of it nearest to the mode of the original type.

and groundwork of the views which he puts forward.

The plant which afforded the material for this discovery is known as *Œnothera Lamarkiana*—that is to say, this is the name of the old species from which the new species were found to be arising. *O. Lamarkiana* is an American plant, but the specimens which de Vries found to be in a state of mutation had made their escape from a garden, and were running wild over a disused potato-field near a town called Hilversum, in Holland. On examining these plants, de Vries found two distinct new forms, which were quite unlike the remainder. Each kind occurred in an isolated patch, as if it had arisen from the seed of a single plant.

No description of either of these forms was to be found in botanical literature, nor were there specimens of them in any of the great herbaria. But when de Vries took seeds from some of the plants and sowed them in his garden, he found that the new forms came true to type — the plants produced resembled the parents from which the seeds were taken, and not the normal form of *O. Lamarkiana*.

Here, then, we have a case in which two new species had originated from an old one in a state of nature. But de Vries went further than this, and took measures for observing the actual origin of new forms in the cultivated offspring of the semi-wild *Œnothera*.

For this purpose he transplanted a number of roots from the field where they were growing, and also took seed from a number of other plants, and from these he cultivated large numbers of seedlings for a series of generations. The net result of his experiments was this : out of about 50,000 individuals which were grown to a recognisable stage, more than 800 showed mutation—that is to say, they differed specifically from the parent *O. Lamarkiana*. The 800 individuals belonged to about fifteen new kinds, most of which appeared repeatedly, though some were more frequent than others. The process of mutation had, therefore, taken place in about 1½ per cent. of the seedlings which were grown, and owing to various reasons this estimate is probably considerably too low. For example, many of the new forms were very weakly, and often died before it was possible to distinguish them. Others, again, could not be recognised until an advanced stage of their growth had been reached, whereas only a small proportion of the seedlings raised could be grown after they had reached any considerable size, owing to considerations of space.

We cannot now follow de Vries very far into his elaborate account of his new species and of the way in which they originated ; a few general remarks only must suffice. Many of the new forms were recognisable as quite young seedlings, notably *O. albida,* others not until a much later period of their growth. *O. gigas* was the finest and strongest of the new forms, but only made its appearance on two occasions. *O. lata* also appeared to be as strong as the parental type, whilst two other forms were able to survive in nature in competition with the original species, as has been already described. Other forms which were grown and flowered were plainly less well fitted for

FIG. 12.—MUTATION IN ŒNOTHERA.

(*From de Vries.*)

Top row ..	..	*Lam.*	*lata*	**Lam.**
Second row	..	*subovata*	*albida*	**Lam.**
Third row ..	..	*albida*	*albida*	*lata*
Fourth row	..	*oblonga*	*lata*	**Lam.**
Fifth row ..	..	**Lam.**	**Lam.**	*rubrinervis.*

[*To face p.* 136.

the battle of life than *O. Lamarkiana*, and only reached the flowering stage by the help of careful cultivation, and others, again, were never got to flower at all. Some of the latter, however, were readily distinguishable by the strikingly original types of radical leaves which they exhibited.

When they had once made their appearance, the majority of the new types came true to seed. Sometimes new mutations appeared among their offspring, but these always appeared in smaller numbers than among the offspring of the parent *O. Lamarkiana*, and some of the commoner mutations were usually omitted, so that it appeared as if the process of mutation was accompanied by a tendency towards a fresh stability. Some of the most marked new forms came quite true so far as the observations were carried.

Speaking generally, the nature of the differences which distinguished the new forms from the parental species was just of the same type as that of those which distinguish Jordan's species when found in nature. The differences were not, as a rule, of the sort shown when new garden varieties arise as sports. An example of this latter kind occurred, however, in the case of the new form *O. nanella*, which was a dwarf or permanently stunted form, but in other respects closely resembled the parent type. Apart from this, the new forms appeared to be given off quite at random, without showing any definite tendency towards progress in a particular direction. One of the new species was almost sterile as far as its ovules were concerned, though producing good pollen, whilst in another the formation of the pollen was very defective. None of the others was lacking in either of these respects. Each new form was distinguished by certain definite features which affected almost all its parts, not by one new character only; and these features were never separable, but always appeared in common on the same plant.

The new species, of course, showed normal fluctuating variability, and, as an extreme result of this variability, forms occasionally appeared midway between one of the new species and the parental type. In such cases, when the self-fertilized seed of the plant showing such an intermediate character was sown, the offspring were found to group themselves round the normal form of the new species *or* round that of the parent *Lamarkiana*, thus affording evidence as to the true nature of their parent.

Whether or not we are prepared to accept the whole of de Vries' conclusions from his experiments, we can see at least that from one point of view they are of the very greatest importance. For before de Vries published this work it had been supposed to be quite impossible to make direct observations upon the manner of origin of new species in Nature. De Vries has now shown that such observations can be made, and this is in itself a most valuable piece of information. He has introduced an entirely new method into the domain of species research, and one by the use of which it is to be hoped that before long a definite answer will be obtained to the question whether species in general arise by definite steps, or with an imperceptible degree of slowness.

When results of the novelty and importance of those which have been published by de Vries are brought to our notice, we are naturally disposed to reserve our acceptance of the conclusions which they seem to indicate until observations have been made in confirmation of them by some competent observer. This has now been done by Professor Macdougal at the New York Botanic Garden. Macdougal has carried out observations similar to those above described upon the offspring of seeds sent by de Vries from Holland, and with closely similar results. Thus he has observed all the new forms which de Vries described, as well as some additional ones ; and he has obtained an even higher percentage of 'mutants' than de Vries himself—namely, about 3 per cent. of the total number of seedlings grown. This last result is probably only due to the application of more thorough methods of investigation, and to a smaller mortality of the weakest plants, arrived at by greater care, and rendered possible by the warmer summer climate and by American efficiency in method. De Vries himself, in one of his later generations, when particular care was applied to the methods of cultivation, obtained nearly 3 per cent. of new forms. Macdougal also states that he has observed undoubted cases of mutation taking place in other species besides *Œnothera Lamarkiana*.

It appears, then, that there can be no doubt about the genuineness of the phenomenon described by de Vries. But it is, of course, quite a different thing to assert that all natural species arise in this fashion, and this is what de Vries' theory, as distinguished from his facts, amounts to. De Vries made observations upon a large proportion of the plants of his district by the method of growing great numbers of their seedlings, but he failed to find the same phenomenon going on in any of them. He therefore supposes that species are subject to comparatively short periods of mutability which recur at relatively long intervals, and that all the species he examined except the _Œnothera_ were in this intermediate staple period of their existence. Direct proof of this suggestion is naturally out of the question.

It will be well to summarize briefly the conclusions at which de Vries has arrived, as the result of his observations upon _Œnothera_.

The following are the points to which he attaches chief importance :

1. The new species arise suddenly at a single step, without transitional forms.

2. They are usually fully constant from the first moment of their origin.

3. The distinctive characters of the new forms agree in kind with those which distinguish from one another such old and established species allied to _Œnothera Lamarkiana_ as _O. biennis_ and _O. muricata_. Only one of the new forms—namely, _O. nanella_, a dwarf type— is analogous with any ordinary kind of variety of garden origin.

4. A considerable number of individuals of the same sort usually make their appearance at the same period.

5. Although the new types vary in a normal fashion, and frequently transgress the limits dividing them

from the parental type, yet their first appearance has nothing to do with normal or continuous variability.

6. The mutations take place indefinitely, showing no special tendency in any particular direction.

7. The tendency to mutate recurs periodically. But, as was previously stated, there is no direct evidence of this last supposition.

In addition to what has already been said with reference to the method of origin of garden varieties in general, de Vries has described a number of special phenomena regarding the behaviour of garden varieties of plants, some of which are of considerable interest. Taken together, the facts substantiate to a great extent the view that selection does not of itself lead to the production of specific characters. But de Vries also introduces certain new conceptions which require to be briefly described on account of their great general interest to practical breeders and gardeners. They consist in the idea of races existing intermediate between a species and a complete variety or sub-type of it. Such *between-races* are of two kinds, of which it is unusual to find both in the case of the same species ; moreover, either of them may occur even when the complete variety is quite unknown. In the case of a *half-race* a small percentage only of seedlings is found to produce plants which show the racial character, the remainder being of the original specific type; and even if the racial type is selected for several generations, the percentage of plants of this type which is produced does not notably increase. A *mid-race*, on the other hand, can readily be improved by selection, and when

best developed as a rule either shows the racial character in about half of the seedlings produced, or else
exhibits in the great majority of its members a combination of the character of the species with that of
the race. As an example, we may take the case of
variegated plants, in which the leaves show streaks
or patches of a yellow colour owing to the want of
development of the proper green tint. An ordinary
variegated plant, then, is looked upon as showing a
combination of the green type with the yellow character of a completely modified race—the *aurea* variety,
although the latter exists as such only in a few rare
cases, in which the plants bear leaves showing no
green pigment at all. On the other hand, many
species of plants produce a small proportion of variegated individuals at each sowing, as is often the case,
for example, with Indian corn ; and this circumstance,
according to de Vries, indicates the existence of the
corresponding half-race.

The relative development of the two coexisting
characters in such cases is highly variable, as anyone
may observe for himself in variegated grasses and
similar plants.

It might be supposed that it would be possible to
pass from the species to the half-race, thence to the
mid-race, and so on to the complete race simply by
selection. De Vries shows that this is very rarely, if
ever, the case. He regards the passage from a half-
race to a mid-race, for example, as a mutation, and his
observations seem to show that this transition is not
more frequent than any other mutations.

FIG. 13.—TRIFOLIUM PRATENSE QUINQUEFOLIUM.
(*From de Vries.*)

[*To face p.* 143.

As a further illustration of what is meant by a between-race, mention may be made of the five-leaved race of purple clover (*Trifolium pratense*) obtained by de Vries, and developed by a process of selection. It would appear that the plants occasionally found growing wild, which bear a single four-lobed leaf, usually belong only to a half-race. De Vries was fortunate enough to find two plants upon each of which several of the leaves showed this anomaly, and from these, by an elaborate process of selection extending over several years, a race was obtained, the leaves of which in the majority of cases showed five lobes, whilst some had six or seven. Since, however, it appeared impossible to get rid of a certain proportion of three-lobed leaves, and equally so, on the other hand, to obtain leaves with more than seven lobes, de Vries concluded that his experiment exemplified the development of a mid-race, and not that of a constant race or true variety.

Some of the experiments upon which the above-mentioned conclusions of de Vries are based are open to the criticism that sufficient precautions do not seem to have been taken to prevent intercrossing between the selected plants and other types of different constitution, and it is only just to mention that such cases are usually pointed out by the author himself. In these cases the possibility of cross-fertilization between the novel type and normal members of the species from which it was derived renders the conditions of experiment closer to those obtaining in Nature ; but full precision in the interpretation of the results is thereby sacrificed, and it will be of interest to have evidence of confirmatory cases in which thoroughly definite methods of pollination have been used.

The views of de Vries with regard to the actual origin of new species may be summed up as follows : Broadly speaking, species arise by mutation, by a sudden step in which either a single character or a whole set of characters together become changed. In the former case a new variety in the strict sense of the word is the result ; in the latter a new species (according to Jordan's definition) is produced.

But mutation may be of several kinds. In the first place, an entirely new character or set of characters may make its appearance. To such a phenomenon de Vries applies the term of *progressive mutation*, and it is by steps of this kind that he believes the main divisions of the vegetable kingdom to have been built up. In the case of such mutations the new character is supposed to come into existence first in a latent or hidden condition, and it may be only after many generations that it makes its appearance visibly. On this view the period of mutation is preceded by a premutation period, during which the appearance of the new character is being prepared for.

A second method of species formation, entitled by de Vries *degressive mutation*, is indicated when a change takes place in the partial latency of a character. A completely latent character is, indeed, unrecognisable as such. But characters may also be only partially latent, and in these cases they exhibit themselves from time to time in rare individuals in the form of sports or abnormalities—a phenomenon which we have already seen to be characteristic of half-races ; indeed, a half-race might have been defined as a strain in which the character of the complete race is usually latent, and only rarely appears. An active character, on the other hand, is apparent in the great majority of the individuals of a race. If, now, a change from latency to activity occurs suddenly, this is a form of mutation. The reverse case, too, may occur—a character previously active may become latent ; the character then appears to be lost, and the mutation is said to be *retrogressive*. De Vries regards the great variety of allied species which is to be found in many groups as being to a large extent the result of retrogressive mutation. This type of mutation is also frequent among cultivated plants. Thus, the appearance of a white variety of a species previously only known to produce coloured flowers may constitute a good example of a retrogressive change. Mutations may also be atavistic, consisting in what is known as a "throw-back" to a previous ancestor. In the most usual form of this phenomenon an ancestral character which had previously become latent shows itself once more in the active condition. Finally, new and distinct types may arise by the intercrossing of separate species, but this is not regarded by de Vries as being an important source of permanent new forms.

Without following de Vries into all the niceties of his theory as to the particular kinds and methods of mutations, we must admit that his experiments go far to establish the doctrine, in support of which a considerable amount of evidence had previously been accumulated, especially by Bateson, that the origin of species in Nature is generally a definite process, and takes place by steps of considerable amplitude. What, then, is the meaning of individual differences, of that continuous variability which is often so considerable, and of the inheritance of this kind of differences which the biometricians have been at so much pains to prove ? De Vries points out that for no two plants are the conditions of life exactly the same ; a considerable degree of diversity among the plants themselves is therefore advantageous, even when these belong to the same specific type. Upon continuous variability depend local races, forms adapted to wetter and drier situations, highland and lowland races, and the like, but none of these are permanent. As regards the cause of this variability, apart from the effect of sexual reproduction, which combines the tendency to vary of two separate parents, de Vries believes that individual variability depends entirely upon nutrition ; but under this head he includes practically the whole environment of plants—light, space, soil, moisture, and the like. Characters acquired in a similar way by previous generations are inherited, and the effect of conditions upon the developing seed whilst still borne upon the parent plant may be considerable. Thus easily does de Vries dispose of the puzzling question of the inheritance or non-inheritance of acquired characters. Acquired characters are inherited ; they are not of any importance in the origin of species.

According to the view upheld by Wallace, Weismann, and others, the actual origin of specific distinctions takes place by natural selection acting upon individual differences ; and in this case it is to be observed that it is the struggle between individuals of the same species which is of primary importance. On the mutation theory it is only the competition between allied species which interests us from the point of view of evolution. Natural selection is thus regarded as having no influence in the formation of species themselves. On the other hand, the gaps existing between genera and still larger groups, such as families and classes, is still supposed to be due to the destructive action of natural selection determining the survival of the fittest species, so that this principle is by no means ousted from its prominent position in the philosophy of evolution even as expounded by the mutationist.

One further point. On the theory of mutation the survival of useless structures becomes readily comprehensible. Indeed, a structure which is actually of the nature of a handicap to its possessor may fail to cause extinction if it is combined with a vigorous constitution, or if it is correlated with other characteristics which are sufficiently useful to make up for the disadvantages entailed. The survival of many apparently useless and some apparently harmful structures is very difficult to understand on the hypothesis of a continuous evolution by the survival of the fittest individuals. This is an argument upon which de Vries lays considerable stress, although it may be pointed out that it is usually very difficult to form a judgment as to the real usefulness or otherwise of organs.

CHAPTER VI

THERE is one side of the practical study of heredity which dates back to the middle of the seventeenth century—namely, that branch of the subject which is concerned with the hybridizing or artificial cross-breeding of different species and varieties of plants. Quite recently the great importance which attaches to this method of study has been realized once more, and the interest thus awakened has led to a closer examination of the accounts of experiments undertaken a century or more ago, with the result of showing that much of the work then carried out in this direction had attained to quite an astonishing degree of excellence. In the brief sketch of the history of hybridizing work here following, account will be taken almost exclusively of experiments of which the interest is not historical only, but which possess an actual scientific value. Amongst other matters of interest, it will be found that more than one observer came very near to anticipating Mendel's epoch-making discovery, and thus arriving at the clue which should unravel almost all the complex problems which beset the early hybridizers.

JOSEPH GOTTLIEB KÖLREUTER, 1733·1806

(*After an engraving by* J. CEDERQUIST.)

[*To face p.* 148.

Following the modern usage, we shall apply the term ' hybrid ' to all individuals arising from a cross bétween parents which belong to distinct groups, no matter whether these groups are separated as distinct genera or species, or whether they are regarded as representing only different races or varieties. This wide interpretation of the term hybrid has only recently been reintroduced. The use to which it has returned is, indeed, the original one ; but many intermediate writers, including Darwin, confined the employment of this expression to cases of crossing between species, and applied the word ' mongrel ' to the offspring of crosses between races or varieties of the same species. Darwin, however, did not regard species as differing in kind from varieties, and he even particularly emphasized the smallness of the distinction which can be drawn between the behaviour and properties of hybrids and mongrels respectively. Indeed, he came to the highly important conclusion that the laws of resemblance between parents and their children are the same, whatever may be the amount of difference between the parents in question—whether, that is to say, they are distinguished only by individual differences, or whether they belong to separate varieties or even species. We have already seen that the more recent facts of biometry point strongly towards the conclusion that individual and race differences are inherited at approximately the same rate. It seems, however, to be at present somewhat doubtful whether all sorts of specific differences follow the same law of propagation on cross-breeding.

Between 1760 and 1766 Joseph Gottleib Kölreuter carried out the first series of systematic experiments in plant hybridization which had ever been undertaken. These experiments not only established with certainty for the first time the fact that the seeds of plants are produced by a sexual process comparable with that known to occur in animals, but also led to a knowledge of the general behaviour of hybrid plants, which was scarcely bettered until Mendel made his observations a century afterwards.

Kölreuter found that the hybrid offspring of two different plants generally took as closely after the plant which yielded the pollen as after that upon which the actual hybrid seed was born. Indeed, he found that it made little or no difference to the appearance of the hybrid which of the parental species was the pollen-parent (male), and which the seed-parent (female)—that is to say, in the case of plants the result of reciprocal crosses is usually identical. Thus, for the first time it was definitely shown that the pollen-grain plays just as important a part in determining the characters of the offspring as does the ovule which the pollen-grain fertilizes. This was a wholly novel idea in Kölreuter's time, and the fact was scarcely credited by his contemporaries.

Kölreuter had no means of discovering that the contents of a single pollen-grain unite with the contents of a single ovule in fertilization. But he ascertained by experiments that more than thirty seeds might be made to ripen by the application of between fifty and sixty pollen-grains to the stigma of a particular flower, so that, if he had had any hint of the actual microscopic processes of fertilization, he would have been quite prepared for the more fundamental discovery.

Kölreuter, indeed, believed that the act of fertilization consisted in the intimate mingling together of two fluids, the one contained in the pollen-grain, and the other secreted by the stigma of the plant. The mingled fluids, he supposed, next passed down the style into the ovary of the plant, and arriving at the unripe ovules, initiated in them those processes which led to the formation of seeds. In this belief Kölreuter simply followed the animal physiologists of his time, who looked upon the process of fertilization in animals as taking place by a similar mingling together of two fluids. Now that we know that fertilization consists essentially in the intimate union of the nuclei of two cells, one of which, in the case of plants, is the ovum contained within the ovule, whilst the other is represented by one of a few cells into which the contents of the pollen-grain divide, we can understand more clearly the bearing of Kölreuter's observation. And it is greatly to this eminent naturalist's credit that he succeeded in carrying out his observations with so much accuracy, when the full meaning of those observations was of necessity hidden from his comprehension.

Kölreuter was the first to observe accurately the different ways in which pollen can be naturally conveyed to the stigma of a flower. This may take place either by the pollen-grains falling directly upon the stigma, or by the agency of the wind, or, lastly, the pollen may be carried by insects visiting the flowers. And he recognised many features characteristic of flowers apt to be fertilized in one or other of these ways in particular. Thus he was aware, for example, of the nature and use of the nectar which so many flowers produce—namely, that it is the substance from which the bees—by far the most diligent visitors of flowers—obtain their honey.

Curiously enough, Kölreuter was not aware of the existence of any natural wild hybrid plants. But he was quite right in contending that supposed examples of such hybrids required for their substantiation the experimental proof, which could only be afforded by making actual artificial crosses between the putative parent species.

The first hybrid made artificially by Kölreuter was obtained in 1760 by applying the pollen of *Nicotiana paniculata* to the stigma of *Nicotiana rustica*. The hybrid offspring of this cross showed a character intermediate between those of the two parent species in almost every measurable or recognisable feature, with a single notable exception. This exception was afforded by the condition of the stamens and of the pollen grains produced by the hybrids. These organs were so badly developed that in all the earlier experiments self-fertilization of the hybrid plants yielded no good seed at all, nor were the pollen grains of the hybrid any more effective when applied to the stigmas of either of the parent species. On the other hand, when pollen from either parent was applied to the stigmas of the hybrid plants, a certain number of seeds capable of germination was obtained, although this number was much smaller than in the case of normal fertilization of either parent species. This partial sterility, affecting in particular the stamens and the pollen which they produce, is a feature common to the majority of hybrids between different natural species. Many such hybrids, indeed, are altogether sterile, so that a further generation cannot in any way be obtained from them. On the other hand, the members of different strains or varieties which have arisen under cultivation yield, as a rule, when crossed together offspring which are perfectly fertile.

In subsequent years Kölreuter was able to obtain a very few self-fertilized offspring from hybrids of the same origin as the above. The resulting plants were described as resembling their hybrid parent so closely as to be practically indistinguishable from it.

The offspring obtained by crossing the hybrid plants with pollen from either parent showed in each case a form more or less intermediate between that of the original hybrid and that of the parent species from which the pollen was derived. But the plants were not all alike in this respect, some of them being much more like the parent species than others, and some, again, varying in other directions. There were also considerable differences between the different individuals in respect of fertility, so that some of the plants were more and some less sterile than the original hybrids. Also, there was some tendency to the production of malformations of the flowers and other parts.

One of the most noted of Kölreuter's experiments was that which consisted in repeatedly recrossing a hybrid plant with one of the parent species from which the hybrid was derived. By continuing to pollinate the members of one generation after another with the pollen of the same parent species, plants were at last arrived at which were indistinguishable from the parent in question. We shall return to this fact later on, when the reader will be in a position to appreciate its importance more fully.

Kölreuter found that the result of reciprocal crosses is usually identical—that is to say, the offspring obtained by fertilizing a plant A with pollen from a plant B are not to be distinguished from those obtained when B is fertilized with the pollen of A. But the two opposite processes of fertilization are not always equally easy to carry out. An extreme instance of this circumstance was met with in the case of the genus *Mirabilis*. *Mirabilis jalapa* was easily fertilized with pollen from *M. longiflora*. During eight years Kölreuter made more than two hundred attempts to effect the reverse cross, but without success.

It was shown by Kölreuter that hybrids between different races or varieties of the same species are usually much more fertile than hybrids obtained by crossing distinct species. Indeed, he believed that varieties of a single species were in all cases perfectly fertile together, whilst hybrids between species always showed some degree of sterility. But in this case Kölreuter based his definition of a species upon the very point at issue, and when he found forms, which other botanists regarded as good species, to be perfectly fertile together, he immediately regarded them as being only varieties of a single species.

One curious point is worth quoting in this connection. Five varieties of *Nicotiana tabacum* were found to be perfectly fertile with one another, but when crossed with *Nicotiana glutinosa* one of them was found to be distinctly less sterile than the rest.

Another interesting point observed by Kölreuter was the fact that hybrid plants often exceed their parents in luxuriance of growth. Upon this fact, as we shall see later on, Knight and afterwards Darwin based theoretical conclusions of considerable importance in connection with the problem of sex.

Thomas Andrew Knight, who was also a botanist of high reputation in other fields, was the earliest observer to lay stress upon the practical aspect of the study of hybrids, and he occupied himself to a considerable extent with the improvement of useful races of plants by cross-breeding. Breeders of animals had already made important improvements by the method of inter-crossing different races, and selecting the most notable types which made their appearance in consequence, when Knight bethought him of applying the same principles to the improvement of plants, and particularly of fruit-trees.

Knight also carried out a series of experiments with domestic peas, the results of which were published in 1779. These experiments have a particular interest from the historical point of view, since it was by dint of similar experiments upon the same kind of plants that Mendel's law was afterwards discovered. This very discovery might even have been made by Knight himself, if he had only realized the importance of ascertaining on a large scale the numerical proportions in which the different kinds of plants, arising in the second generation from the crosses, made their appearance. Unfortunately, this particular form of inquiry never seems to have occurred to him.

Knight's experiments were made with a different object in view—namely, that of discovering whether a cross with a distinct race would provide the stimulus necessary to restore its lost vigour to a strain of plants which was supposed to have become debilitated, owing to its members having been bred exclusively by self-pollination for a long series of generations.

The result of the experiments undoubtedly established the fact that in some cases the hybrid offspring of two distinct races shows a more vigorous habit of growth than either of the parental types. The following extract from Knight's own account will indicate the nature of the experiments upon which his conclusions rest :

' By introducing the farina of the largest and most luxuriant kinds into the blossoms of the most diminutive, and by reversing this process, I found that the powers of the male and female, in their effects on the offspring, are exactly equal. The vigour of the growth, the size of the seeds produced, and the season of maturity, were the same though the one was a very early, the other a very late variety. I had in this experiment a striking instance of the stimulative effects of crossing the breeds, for the smallest variety, whose height rarely exceeded two feet, was increased to six feet, whilst the height of the large and luxuriant kind was very little diminished.'

We shall see, however, that the phenomenon last alluded to admits of a different interpretation.

It was upon the somewhat slender basis afforded by this experiment that the generalization known as the Knight-Darwin law was originally established. Knight's own expression of this idea was to the effect that 'Nature intended that a sexual intercourse should take place between neighbouring plants of the same species.' And the same conclusion was expressed still more forcibly by Darwin in the aphorism : ' Nature abhors perpetual self-fertilization.' But although it may be true that in a considerable number of cases advantages are gained from the process of cross-fertilization between different members of the same species, which do not accrue when self-fertilization takes place, yet several cases are now known in which self-fertilization really does seem to be indefinitely continued.

Knight crossed a pea having white flowers and seed-coats, and green stems, with one in which the flowers and stems were coloured purple, and the seeds grey. The seeds immediately resulting from the cross were unchanged in appearance, but the plants arising from these took closely after their coloured male parent. On crossing the cross-bred plants once more with a white strain a certain proportion of white plants was again obtained, though what that proportion was

Knight failed to notice. He observed, however, that white crossed by a purple strain invariably gave purple, whilst the cross-bred purples, when crossed again with white, yielded some white and some purple plants.

In 1822 John Goss recorded the fact that a ' blue ' pea crossed with a ' white ' yielded from the crossed flowers pods with white seeds only, the seeds contained in other pods upon the same plant being, of course, blue. The plants produced from the white seeds bore some pods with all blue, some with all white, and many pods with both white seeds and blue ones ; and a coloured plate is given which shows one of the latter pods together with its contents. The blue seeds, when sown separately, yielded plants which produced blue seeds only, but plants arising from the white seeds yielded a mixture of blue and white seeds.

Knight pointed out quite correctly that the colours of the seeds which are here referred to are occasioned by the colour of the cotyledons or seed-leaves of the pea, which are visible through the semitransparent seed-coat. Green cotyledons give rise in this way to a bluish appearance, whilst, when the cotyledons are yellow, the resulting appearance of the seed is described as whitish.

The Hon. and Rev. W Herbert was another observer who made many important experiments in hybridization towards the beginning of the nineteenth century. These led him to the conclusion that Kölreuter and Knight were wrong in their assertion that hybrids between distinct species were always sterile. Herbert considered that only generic or family types were constantly sterile, and this led him to the further conclusion, now believed to be erroneous, that the separate genera or families were those which were originally created, whilst he believed that the separate species of the same genus arose from a single original type by a genuine process of evolution.

The most prolific in work of all the hybridists, however, was undoubtedly Carl Friedrich v. Gaertner (1772 - 1850). Gaertner made a great number of crosses between species belonging to all sections of the natural system, and his book, published in 1849, contains a great mass of valuable information. Gaertner's theoretical conclusions, for the most part, only amplify and confirm those of Kölreuter, upon whom in this direction he made but little advance.

C. Naudin's essay, entitled 'New Researches on Hybridity in Plants,' made its appearance in 1862. The author pointed out that the facts of the return of hybrids to the specific forms of their parents, when repeatedly crossed with the latter, are naturally explained by the hypothesis of the disjunction of the two specific essences in the pollen grains and ovules of the hybrid. The idea may, perhaps, be made somewhat clearer as follows : Let us consider the case of a species A crossed with another species B. Naudin supposes that some of the pollen grains and ovules of the hybrid plant will be potentially* of the exact

* When it is said that a pollen grain or ovule potentially resembles the species A, it is meant that the germ-cell in question is of such a kind that, when united with one derived from an ovule or pollen grain of similar constitution, it would give rise to a plant exactly resembling A.

character of one species (A), whilst others will bear no potential resemblance to A, but will be precisely similar in nature to the ovules and pollen grains of the pure species B. In cases where this separation of the materials representing the two types in a potential condition is complete, forms exactly resembling the parents might be obtained. As we shall see, this hypothesis makes a remarkably near approach to that of Mendel ; and the importance of the fact that the first hybrid generation is generally uniform, as contrasted with the diversity of types often appearing in the second generation, is clearly recognised by Naudin. This observer considered the hybrid in the adult state to consist of an aggregate of particles, homogeneous and characteristic of a single species when taken separately, but mingled in various proportions in the organs of the hybrid, which is thus looked upon as a kind of living mosaic.

The only other discovery of first-class importance,* in addition to that of Mendel, made during the nineteenth century in the domain of hybridization, was that published by Millardet in 1894.

Millardet's principal experiments were made upon strawberries, of which plants he crossed together a number of different species and varieties. Contrary to what had been observed in the majority of such crosses between other species of plants, in which the offspring was usually more or less intermediate between the two parents from which it arose, Millardet

* That is to say, if it is really genuine. The phenomena do not appear to have been seen by anyone else.

found in a considerable number of cases that the offspring resembled one parent only, from which it was indeed indistinguishable, whilst no trace of likeness to the second parent could be detected in it. In some cases the resemblance was to the paternal species (pollen-parent), and in others to the maternal species (seed-parent). In several instances the hybrid offspring, on being self-fertilized, bred true to the type which they already exhibited, so that the second generation, like the first, seemed to derive its whole constitution from one parent, to the total exclusion of the other.

The precise meaning of this remarkable phenomenon is not clearly understood. There is some doubt as to whether Millardet's experiments were really sufficient to establish it as a scientific fact. Moreover, Millardet's observations have never been confirmed by later workers. In the absence of directly contradictory evidence it seemed necessary to draw attention to the facts as they have been described.

Great numbers of observations upon the characteristics and behaviour of hybrid plants and animals have been from time to time recorded, and the preceding pages contain only a brief selection of such facts as are most necessary for a proper understanding of modern work in hybridization. Until quite recently the laws of transmission of characters in hybrids were still completely hidden. The facts were wonderful enough, but they showed no signs of falling into an orderly arrangement. In the next chapter it will be our business to describe the remarkable discovery which has introduced order into this previously chaotic region, and which has enabled a few workers to establish in half a dozen years the foundations of a great science, the importance of which is not at all generally realized.

GREGOR JOHANN MENDEL.

[To face p. 163.

CHAPTER VII

WE have already had occasion to point out how important it is, when engaged upon questions of heredity, not to treat whole animals or plants as units, but to deal with their separate characters one at a time. In the course of the present chapter the reason for proceeding in this way will appear more clearly, and we shall find that the adoption of this method is fully justified by the results which it enables us to obtain, and which could not have been arrived at in any other way. We shall also find reasons for believing that this method is the correct one from a theoretical point of view.

Naturally, considerable care is necessary in determining what are and what are not separable characters. At the outset it is not always possible to make this discrimination with certainty, but during the course of the experiments which follow it is almost always possible to arrive at a clear definition of each character, and in many cases the distinction of characters is quite obvious from the beginning.

Up to the present time the experimental study of heredity by the methods of definite breeding has yielded clear and definite information only when applied to cases where clearly definable characters have distinguished the parental forms examined. This, however, is in great part due to the fact that the experimental method has scarcely yet been used in dealing with characters of a less definite nature. The science is still in its infancy, and attention has naturally been first paid to the simpler problems which it affords. The difficulties of treatment which confront those who would deal with highly variable characters and those of a ' more or less ' nature are considerable, although there is no reason for supposing that such problems are insuperable. As we have seen, however, the majority of characters which distinguish species or races from one another appear to be of a perfectly definite description, so that the limitation just referred to is not so serious as might appear at first sight. The recent revival of work upon the subject of inheritance by the use of breeding methods has, as a matter of fact, already been rewarded with results as valuable and as clear as could possibly have been anticipated— results which are sufficient in themselves to show that the discovery made by Mendel was of an importance little inferior to those of a Newton or a Dalton.

It is important to remember that every animal or plant, which has come into existence in the ordinary way through sexual generation, owes its individuality to the mingled natures of two separate parents. The following lines, in which the poet Goethe speaks of his own hereditary endowment, have been quoted more than once in this connection :

' Vom Vater hab' ich die Statur,
Des Lebens ernstes Fuhren,
Vom Mutterchen die Frohnatur
Und Lust zu fabulieren.'

In such a case we must always look upon the corresponding character of the second parent as existing in the offspring side by side with the character which finds expression, only the former is overpowered by the latter, and forced to remain invisible. That the hidden character is nevertheless actually present is shown by the fact that a feature characteristic of a particular grandparent, which did not appear in the parent, may reappear in the child. For instance, a characteristic masculine feature of the maternal grandfather may thus make its appearance in the son.

It is found that any individual may be looked upon as being to a large extent an aggregation of separate characteristics. In a pair of allied races a great number of the separate characters are the same in the two cases, the distinction between the two forms depending upon a few definite features only. The majority of salient characteristics are identical in such a pair, but some of the corresponding characters are opposed. Thus in different races of mankind complexions may be dark or fair, eyes blue or brown, hair straight or curly, and the like. Now the offspring of parents which had smooth and curly hair respectively might exhibit smooth or curly or intermediate (wavy) hair, according as one or the other character, or both in combination, made its presence obvious ; for in the simplest cases both will necessarily be present, though one may be hidden. What will happen in the grandchildren ?

The manner in which characters comparable with the above are actually transmitted has been worked out in the case of many races of animals and plants, and in cases where experimental matings can be readily carried out, and a large number of offspring reared, it is found that a simple rule applies which holds good in every example thoroughly examined hitherto. This law was discovered by Mendel about the year 1865, and has since been called by his name. Before enunciating it we shall consider the information afforded by the case of a single pair of simple characters. Afterwards we shall endeavour to show the application of the law to the more complex cases in which combinations of characters are concerned.

A grain of Indian corn or maize contains a germ or *embryo*, which under suitable conditions will give rise to the future plant. The embryo is surrounded by a certain amount of reserve food material constituting the *endosperm*—a store which is made use of by the young plant during its germination. The embryo arises as the result of a process of fertilization which takes place in the following manner : The ovum, or female cell hidden in a flower, contains a nucleus, and this on fusion with one of the nuclei derived from a grain of pollen initiates the vital processes which lead to the development of an embryo plant.

Nuclei are the central and, from the point of view of heredity, the most important parts of cells—the constituent units of the plant body. The cells which, together with their nuclei, take part in the process of fertilization are known as *gametes*, or *germ-cells*—male and female respectively, the latter being the ovum.

It is less generally known that the endosperm of a grain of Indian corn arises by a very similar process to the one which gives rise to the embryo itself. A second nucleus derived from the same pollen grain fuses with a nucleus situated near the ovum, and to the product of this fusion the endosperm owes its origin. It is further found, so far at least as those characters are concerned to which we shall at present confine our attention, that these two important nuclei hidden in the same female flower are exactly alike in hereditary constitution, and so are the two generative nuclei derived from a single pollen grain. In consequence of this fact, the observed character of the endosperm may be regarded as a true guide to the nature of the plant into which the associated embryo will afterwards develop. The hereditary qualities of the two are exactly the same.

It is not difficult to find a variety of Indian corn in which the endosperm is yellow, and another in which the colour of this tissue is white, owing to the absence of any visible yellow pigment. If a female flower of a white variety is fertilized with pollen taken from a yellow variety, the resulting grain shows its hybrid nature by the presence of the yellow colour in its endosperm. This is found to be a regular rule. Grains upon a plant of a white strain which has been pollinated with ' white pollen ' are white, but if pollinated from a yellow strain the grains are yellow. On the other hand, the grains upon a plant belonging to a yellow strain retain their yellow colour even if the flowers which produce them have been pollinated from a white variety.

These facts are expressed in technical language by saying that yellowness is *dominant* over whiteness, and the latter is said to be *recessive*.

Let us now suppose that we have sown a number of the yellow grains derived from the cross *yellow* × *white** or *white* × *yellow*, and that we have exposed the female flowers of the resulting plants at the proper stage of their existence to the influence of pollen derived from a pure white strain, taking care that none of their own hybrid pollen falls upon them at the same time. The result of this experiment takes us at once to the very heart of the Mendelian theory. Half the total number of grains obtained in this way —from the cross (*white* × *yellow*) × *white*—are white, and half are yellow.

Thus in an experiment carried out in the manner described there were obtained upon ninety-five plants :

> Yellow grains 26,792, or 50·03 per cent.
> White grains 26,751, „ 49·97 „

But we must go further than this. On sowing the white grains obtained in this second generation (F_2), and allowing the plants obtained from them mutually to pollinate one another, cobs were obtained bearing exclusively white grains without any trace of yellowness.

* × is to be read ' fertilized with pollen from.

Half the grains, then, of the parentage (*white* × *yellow*) × *white* are pure white in colour, and not to be distinguished from grains of the parentage *white* × *white* even after an extensive examination of their offspring, which is the most rigorous test we are able to apply.

The yellow grains born upon the same hybrid plants (F_2) had clearly each of them one white parent—namely, the plant from which the white pollen was derived. On sowing these yellow grains and once more pollinating by pure white, a precisely similar result was observed to that obtained in the preceding generation—that is to say, these plants, derived from yellow grains, produced once more 50 per cent. of white grains and 50 per cent. of yellow. We are, therefore, led to suppose that the yellow grains born upon the hybrid plants are of precisely the same nature as the original yellow hybrid grains (*white* × *yellow*), since their behaviour when pollinated from the same white strain is identical. We may express the result so far obtained in the form of the following diagram :

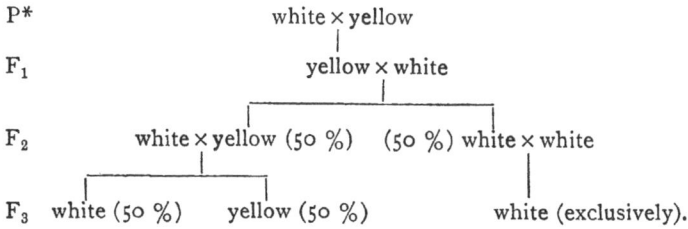

P*   white × yellow

F_1 yellow × white

F_2 white × yellow (50 %)   (50 %) white × white

F_3 white (50 %)   yellow (50 %)   white (exclusively).

* The following shorthand expressions are adopted to denote the different generations in cross-breeding experiments :   P is the generation of the original parents ; F_1 is the first generation of offspring—the cross-bred seeds and the plants to which they give rise. To the F_2 generation belong the seeds produced upon the F_1 plants, and the plants to which they give rise, and so on.

The pollen of the F_1 plants (*i.e.*, those plants which were derived from the yellow cross-bred grains)—when applied to the female flowers of the same pure white strain of maize, caused in like manner the appearance of white and yellow grains in equal numbers. This result is equally well expressed by the above diagram on simply regarding the yellow in F_1 as the male parent (pollen-parent) instead of as the female parent (seed-parent) of F_2.

What is, then, the meaning of these results ? The case is really very simple. The germ-cells (ova and pollen-nuclei) of the cross-bred plants (*white × yellow*) must be potentially either pure white or pure yellow, with no blending of these characters. Further, the two kinds (yellow and white) of male germ-cells or pollen-nuclei must arise in equal numbers, and the same must be true of the female germ-cells or ova. By this supposition only can the observed facts be explained. If the supposition is true, then, when the cross-bred plant (F_1) is crossed again with the pure white form, its white germ-cells give rise to white grains which are of the nature (*white × white*), and are therefore pure. Its yellow germ-cells give rise to yellow grains which are of the nature (*yellow × white*). And, since the number of yellow- and white-bearing germ-cells is equal, the number of yellow and of white grains produced in this way is approximately the same. The yellow grains are of the same composition as the original cross-bred grains obtained by crossing pure white with pure yellow, and we have seen that they behave in exactly the same way on further cross-breeding. This conclusion is at least so far firmly established that no alternative hypothesis has been put forward which will explain the facts.

We have next to consider what will be the result of crossing our cross-bred plants with one another instead of with the pure white form. The following possibilities present themselves :

A yellow female gamete may pair with a yellow male gamete.
  ,,   ,,   ,,   ,,   ,, white ,,
A white ,,   ,,   ,,   ,, yellow ,,
  ,,   ,,   ,,   ,,   ,, white ,,

All these combinations are equally likely to occur, because in each plant there are the same number of yellow and white female gametes as well as of yellow and white male gametes. In the long-run, therefore, each of the above pairings will be found to have taken place in an equal number of cases. The grains which we shall obtain, then, will be yellow and white in colour, and the two kinds will occur in the following proportions : 1 pure white ; 2 white × yellow or yellow × white, which, as we have already seen, will be yellow in appearance ; and 1 pure yellow. Altogether, we shall expect a ratio of 3 yellow grains to 1 white.

In an actual experiment the following result was obtained :

Yellow grains 16,592, or 74·5 per cent.
White   ,, 5,681, ,, 25·5 ,,

—that is to say, a ratio of 2·9 yellow to 1 white.

The expression $1A : 2Aa : 1a$, in which A represents the dominant character (yellow) and a the recessive character (white), may be spoken of as a Mendelian formula. It indicates the proportion in which the two pure types and their hybrid brethren will appear, on breeding together the offspring of a simple or *mono-hybrid* cross—*i.e.*, one in which attention is paid to the behaviour of a single pair of characters only.

So far we have been dealing with a pair of characters consisting in the presence and absence respectively of a particular pigment. Precisely similar results are to be obtained in the case of a pair of *structural* characters. The endosperm, or reserve substance, of certain varieties of Indian corn shows a smooth surface, and contains an essentially starchy reserve material, whilst in other races the surface of the endosperm is wrinkled and the reserve product is of a sugary nature. This sugary endosperm is characteristic of the kinds of corn largely used in the United States of America as a table vegetable.

On crossing together a variety with smooth starchy grains and one with wrinkled sugary grains, the grains immediately resulting are smooth and starchy, no matter whether the starchy strain is used as the seed-parent or as the pollen-parent—that is to say, the starchy character is dominant, a dominant character being one which appears in F_1 to the complete or almost complete exclusion of the corresponding character exhibited by the other parent, which is spoken of as recessive. In the present case the sugary character is recessive.

The further behaviour of the cross between smooth and wrinkled is precisely the same as that of yellow crossed with white. Thus, if the hybrid plants are bred together or self-fertilized, the resulting cobs will exhibit a proportion of three smooth grains to one wrinkled grain.   In an actual example there were obtained 5,310 smooth grains and 1,765 wrinkled, or 75·06 per cent. of the former and 24·94 per cent. of the latter.

In a further generation the wrinkled grains breed true.   One out of every three smooth grains does the like.   The remaining two smooth grains are of hybrid nature, and on self-fertilization yield again the same proportion of three smooth to one wrinkled.   Such hybrid grains and the plants into which they develop are spoken of as *heterozygotes*.

Thus, if we write *B* for smooth and *b* for wrinkled, the following scheme will express the result of crossing together plants which bear these characters, and afterwards self-fertilizing the offspring obtained :

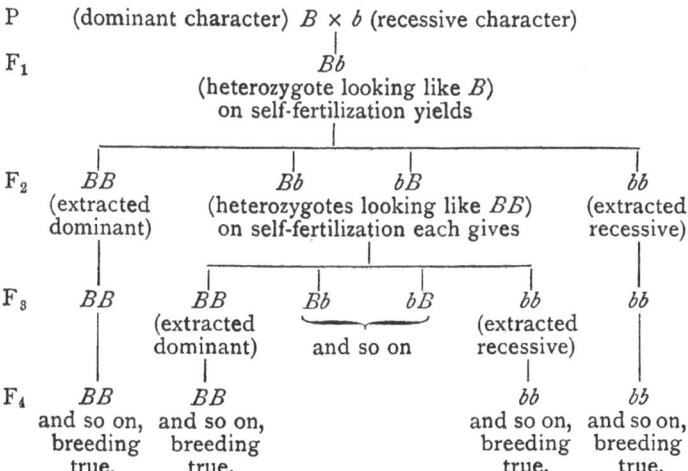

P   (dominant character) $B \times b$ (recessive character)

F_1 *Bb*
            (heterozygote looking like *B*)
                on self-fertilization yields

F_2 *BB*   *Bb*   *bB*   *bb*
      (extracted   (heterozygotes looking like *BB*)   (extracted
      dominant)   on self-fertilization each gives   recessive)

F_3 *BB*   *BB*   *Bb*   *bB*   *bb*   *bb*
                 (extracted   (extracted
                 dominant)   and so on   recessive)

F_4 *BB*   *BB*   *bb*   *bb*
      and so on, and so on,   and so on, and so on,
      breeding   breeding   breeding   breeding
      true.   true.   true.   true.

So far we have seen that both a pair of structural characters and a pair of colour characters can 'Mendelize,' according to the phrase coined by the Germans—that is to say, the germinal representatives of such pairs of characters remain perfectly distinct in the hybrid plant, and separate completely at the formation of its gametes, in such a way that an equal number of gametes arises containing either character.

The members of a pair of characters which behave in this way on crossing are called *allelomorphs*. When a pair of gametes fuse together in the process of fertilization the resulting cell is known as a *zygote*. A zygote formed by the conjunction of two like gametes is called a *homozygote*. When the gametes contain opposite members of a pair of allelomorphs the result is called a *heterozygote*. The same terms may also be applied to the adult multicellular organisms into which these fertilized egg-cells develop.

We have still to consider what happens when parents are crossed which differ in more than one pair of allelomorphs. The actual result is as follows :

Suppose a smooth yellow type of maize to be crossed with a wrinkled white variety, both smoothness and yellowness being dominant. The grains produced in F_1 are therefore yellow and smooth. Let the F_1 plants, arising from the smooth yellow heterozygote grains, be crossed with the wrinkled white parent, which is recessive in respect of both these characters. In this way the true nature of every germ cell produced by the heterozygote will be able to manifest

FIG. 14.—MENDELIAN PROPORTIONS IN MAIZE.

Cobs born by heterozygote plants pollinated with the recessive, showing
equality of smooth and wrinkled and of coloured and white grains.

[*To face p.* 175.

itself in the visible character of the grain produced from it.

The following result was actually obtained in this way :

Smooth yellow grains 2,869, or 25·3 per cent.
Smooth white grains 2,933, or 25·7 ,,
Wrinkled yellow grains 2,798, or 24·5 ,,
Wrinkled white grains 2,803, or 24·5 ,,

Thus we see that a nearly equal number of the germ cells of the double heterozygote bears each of the four possible combinations of characters—that is to say, it is an even chance whether a particular gamete, which bears the allelomorph yellowness, bears also smoothness or wrinkledness. In other words, the two pairs of allelomorphs segregate in entire independence the one of the other. It is particularly to be noticed that we arrive in this way at two perfectly new combinations of characters, which were not shown by the original parent strains. We have synthesized two new sorts of maize with smooth white and wrinkled yellow grains respectively. In a precisely similar way, if the cross is made between strains of which the grains are respectively smooth white and wrinkled yellow, we should obtain in F₂ the new combinations smooth yellow and wrinkled white.

The result obtained on self-fertilizing the hybrid plant is somewhat more complicated.

If we write A for yellowness, a for whiteness, B for smoothness, and b for wrinkledness as before, $AB \times ab$ gives the heterozygote $ABab$. Equal numbers of the germ cells of the heterozygote will be of the compositions AB, Ab, aB, and ab.

All the following zygotic combinations are, then, equally likely :

$$ABAB \quad ABAb \quad ABaB \quad ABab$$

$$AbAB \quad AbAb \quad AbaB \quad Abab$$

$$aBAB \quad aBAb \quad aBaB \quad aBab$$

$$abAB \quad abAb \quad abaB \quad abab$$

Altogether there are sixteen combinations. The result can be expressed more shortly in the form $(A + 2Aa + a)(B + 2Bb + b)$,* which will be found to give the above terms when expanded. Thus the combination of the Mendelian formulæ for F_2 when each of the pairs of allelomorphs is considered separately, gives us the formula for the two pairs of allelomorphs considered simultaneously.

The same result may also be written in the form :

$$\left.\begin{array}{l} 4\ A \\ 8\ Aa \\ 4\ a \end{array}\right\} \text{combined with} \left\{\begin{array}{lll} 1\ B & 2\ Bb & 1\ b \\ 2\ B & 4\ Bb & 2\ b \\ 1\ B & 2\ Bb & 1\ b \end{array}\right.$$

or

AB	$2\ ABb$	Ab
$2\ Aab$	$4\ AaBb$	$2\ Aab$
aB	$2\ aBb$	ab *

* It is customary to condense these expressions as far as possible by never repeating the same letter more than once in each term. Thus, A stands for AA, b for BB, and so on. On expansion, *i.e.*, multiplying together the contents of the two brackets, $A \times B$ gives $ABAB$, $A \times Bb$ gives $ABAb$, and so on for all the other terms of the expression.

Let us consider the external appearance of these various types in the particular example before us.

Nine of the above sixteen terms include A and B, and are therefore smooth yellow in appearance. (We need not stop to consider whether a or b or both are present in addition, since these are recessive.)

Three terms include A and b, B being absent. These, therefore, appear wrinkled yellow.

Three include a and B, A being absent. These, therefore, appear smooth white.

One contains a and b only, and is, therefore, wrinkled white.

With regard to internal constitution :

The nine individuals of appearance AB include the following types :

> One pure, $ABAB$, breeding true to the smooth yellow type on self-fertilization.
>
> Two $ABAb$, heterozygous in respect of the pair B-b, but pure yellow.
>
> Two $ABaB$, heterozygous in respect of A-a, but pure smooth.
>
> Four $ABab$, heterozygous in respect of both pairs of characters.

The three individuals of appearance Ab include the following types :

> One pure, $AbAb$, breeding true to the (new) wrinkled yellow type.
>
> Two $Abab$, giving both wrinkled yellow and wrinkled white.

The three individuals of appearance aB include the following types :

One pure, *aBaB*, breeding true to the (new) smooth white type.

Two *aBab*, giving both smooth white and wrinkled white.

The remaining individual is *ab* in appearance and *abab* in constitution, and breeds true to the wrinkled white type.

The expected behaviour of all these different types can be followed out by the aid of suitable breeding experiments, and not only has this been done in the case of the cross which we have been considering, but precisely similar phenomena have been shown to be taking place in a large number of other characters in many different species of plants and in a good many animals as well.

We are now in a position to state the important proposition known as Mendel's law, which is to the following effect :

The gametes of a heterozygote bear the pure parental allelomorphs completely separated from one another, and the numerical distribution of the separate allelomorphs in the gametes is such that all possible combinations of them are present in approximately equal numbers. (Note that it is impossible for both members of the same pair of allelomorphs to occur together in the same gamete.)

This is the essence of the great discovery made by Gregor Mendel, Abbot of Brunn, and published by him in the Transactions of the Brunn Natural History Society in 1866. By some extraordinary chance Mendel's paper was entirely lost sight of until the same facts were independently rediscovered in 1899 by de Vries working in Holland, by Correns in Germany, and by Tschermak in Austria.

Gregor Johann Mendel was born on July 22, 1822, at Heinzendorf, near Odrau, in Austrian Silesia. In 1843 he entered as a novice the Augustine Convent at Altbrunn, and was ordained priest in 1847.

Mendel was a teacher of natural science in the Brunn Realschule from 1853 to 1868, when he was appointed Abbot of his monastery. During this time he was largely occupied with experiments in cross-breeding a great variety of plants, and some idea of his activity in this line of scientific work is to be gathered from a perusal of his letters to the German biologist Nägeli, a correspondence which has recently been published by Professor Correns. Mendel himself only published the result of his work with peas, and that of a few of his experiments with *Hieracium*.

After 1873 the cares associated with the position of Abbot of Brunn appear to have prevented further biological work. His death took place in 1884, two years after that of Charles Darwin, to whom Mendel was thirteen years junior.

Mendel's own experiments—that is to say, the chief ones published by him—were made with peas, a kind of plants which were found to be remarkably well suited to this kind of work. Seven pairs of characters in these plants were found to behave in precisely the same manner as those characters of the maize-plant which have already been described, and in all of them the phenomenon of dominance also appeared. The characters dealt with by Mendel were as follows, the dominant member of the pair being in each case placed first :

Smooth seeds, and wrinkled seeds.

Yellow, and green reserve material—*i.e.*, cotyledons.

Deeply coloured (grey), and nearly colourless testas or seed-coats.

Inflated or stiff, and wrinkled or soft pods.

Green, and yellow pods.

Flowers scattered up the stem, and flowers in a terminal bunch or umbel.

Tall, and dwarf stems.

As the result of these experiments Mendel came to the conclusion with which his name is now closely associated—that the male and female germ-cells of hybrid plants contain each of them one or the other member only of any pair of differentiating characters exhibited by the parents, and that each member of such a pair of characters is represented in an equal number of germ-cells of both sexes. Furthermore, separate pairs of differentiating characters (allelomorphs) conform to this law in complete independence of one another.

Although in Mendel's own experiments one member of each pair of differentiating characters was always dominant, dominance is by no means an universal phenomenon when different varieties of plants are crossed together. In a considerable number of instances the heterozygote is found to exhibit an appearance which is more or less intermediate between the types of character shown by the parents. It may be almost exactly intermediate, or the appearance of the cross-bred form may be nearer to that of one parent than to that of the other. Dominance is clearly only an extreme case of this latter phenomenon. The term 'dominance' is applied to those cases in which the appearance of the hybrid offspring is so near to that of one parent as to be no longer clearly distinguishable from it.

In other cases, still of a simple Mendelian nature, the appearance of the heterozygote may be quite different from that of either parent homozygote. An excellent example which is almost certainly of this nature is afforded by the Andalusian fowls studied by Messrs. Bateson and Punnett. And this will also serve as our first illustration of the application of these principles to animals as well as to plants. The facts of the case are as follows :

The 'blue' type of Andalusian appears to be a heterozygote form which has never been got to breed true. When a pair of these birds are mated together only about half their offspring are like themselves, the remainder being entirely different. Half these remaining 'wasters' are black, and half are nearly white, showing only a few black 'splashes.' If, now, a pair of the black wasters are mated together, they breed perfectly true, yielding only black offspring like themselves. Similarly the splashed whites mated together give rise to splashed white, and nothing else. Both these forms, then, the black and the splashed white, are clearly pure homozygotes. On mating a black and a splashed white together, black-bearing gametes and white-bearing gametes will meet together in fertilization. In every case in which this form of mating was carried out the resulting chicks were invariably blue.

The gametes of the blue type of Andalusians, then, according to our supposition, do not bear the blue character at all. Half of them contain the black and half of them the splashed white allelomorph. Such gametes, meeting by chance when a pair of blue Andalusians are mated together, give rise to the zygotes —one black-black, two black-white, one white-white— the black-whites being, of course, blue in appearance as before.

Now, we may put this explanation to the test by a very simple experiment—namely, by mating the supposed heterozygote blues with the black and with the splashed white types respectively. Both these forms of mating were examined by Bateson and Punnett, and the results were as follows : It was found that blues crossed with blacks gave rise to equal numbers of blue and of black offspring, whilst when blues were crossed with splashed whites there appeared blue and splashed white chicks in equal numbers. And by a repetition of the process it could be shown that the blues so obtained were heterozygotes as before. Here, then, we have clear evidence that equal numbers of the germ-cells produced by the blue birds bear the pure black allelomorph and the pure splashed white allelomorph respectively, since half the offspring obtained on mating the blue birds with black are black, and half the offspring obtained on mating them with

FIG. 15.—PRIMULA SINENSIS CROSSED WITH P. STELLATA.

Above, the parents. In the middle, the heterozygote offspring
—*P. pyramidalis.* Below, the result of self-pollinating *P. pyramidalis* :
1. *P. sinensis :* 2. *P. pyramidalis :* 1. *P. stellata.*

[*To face p.* 183.

splashed white are splashed white. The following scheme of inheritance illustrates the phenomena described :

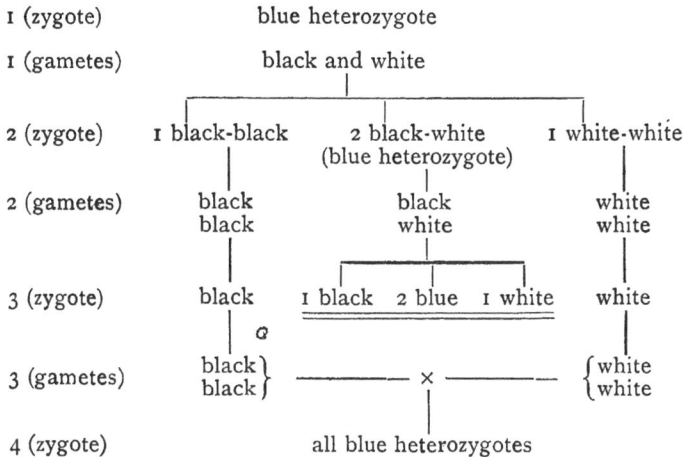

1 (zygote)	blue heterozygote		
1 (gametes)	black and white		
2 (zygote)	1 black-black	2 black-white (blue heterozygote)	1 white-white
2 (gametes)	black black	black white	white white
3 (zygote)	black	1 black 2 blue 1 white	white
3 (gametes)	black} black}	———— × ————	{white {white
4 (zygote)	all blue heterozygotes		

A case which is closely similar to that of the Andalusian fowl is afforded by the cross between *Primula sinensis* and *Primula stellata*.

P. sinensis crossed with *P. stellata* gives rise to a type which is different from either parent, being in some respects intermediate between the two. The hybrid is so distinct that a special name has been given to it, and the new type is known as *P. pyramidalis*. So far it has been found impossible to obtain a strain of *P. pyramidalis* which will breed true. On self-fertilization the offspring are found to show the types of *P. sinensis*, *P. pyramidalis*, and *P. stellata* in the ratio of 1 : 2 : 1.

Cases like the above illustrate the essential part of

Mendel's law even better than those in which dominance is present, the characteristic proportion of one of each homozygote type to two of the heterozygote being at once recognisable in such a case without the necessity for further breeding; whereas, in cases where there is dominance, further study is necessary in order to distinguish, among the individuals of dominant appearance, those which are pure dominant and those which are heterozygous in constitution.

In concluding our account of the simpler forms of Mendelian phenomena we may consider one further point with regard to the nature of the two allelomorphs making up any particular pair. In what is probably a majority of the cases hitherto examined the dominant and recessive allelomorph seem to represent respectively the presence and absence of something. Thus the dominance of colour to absence of colour, or whiteness, is a very frequent phenomenon. And in some of the more complex cases to be described in the next chapter we shall find the presence and absence of a particular factor very often behaving as a pair of Mendelian allelomorphs. The question arises as to how far this conception should be extended. It seems, for instance, somewhat far-fetched to speak of dwarfness as being simply determined by the absence of the factor for tallness, though it is not impossible that this may be the correct way of looking at the facts. Be this as it may, it is to be remembered that a Mendelian pair often represents the presence and absence respectively of a particular feature.

CHAPTER VIII

MENDELISM (*continued*)

MENDEL'S law, as stated in the preceding chapter, has already been found to hold good in a very large number of cases—cases in which all kinds of characters are concerned, belonging to many different species of animals and plants. In certain instances, however, complications arise, and these may be treated of in two main sections.

The first kind of complication arises from the phenomenon known as *coupling*. The essence of this phenomenon consists in the existence of some kind of affinity occurring in the same individual between allelomorphs which belong to distinct pairs. In consequence of such an affinity exceptions are found to the rule that separate pairs of allelomorphs segregate independently.

The closeness of the connection between the characters concerned shows a series of gradations in different cases. In the simplest cases of all, what are loosely spoken of as separate characters are found on closer examination to be only different aspects of one and the same characteristic feature. These cases, then, offer no real exception to the rule, for only one pair of allelomorphs is actually concerned. As an example, we

may take the case of the wrinkled sugary type of maize
already contrasted with the smooth starchy variety.
The essential difference between the two kinds depends
upon the fact that in the former the reserve product
laid down in the endosperm is different, being largely
of a sugary nature instead of being starchy.   With
this circumstance is associated the presence of a larger
proportion of water in the unripe grain. And the
result of this is that, when the grain dries, its surface
falls into folds. The sugary nature of the grains also
causes them to take on a more hyaline or semitransparent appearance than the grains of the starchy
variety.   All these characters, if they can be so called,
behave on crossing as a single Mendelian allelomorph,
and are doubtless represented in the germ cells by a
single substantive representative.

A simple example of what may probably be regarded
as a real case of coupling is afforded by certain colour
characters exhibited by pea-plants. In these plants
coloured flowers, a red or purple colouration in the
axils of the leaves, and a marked pigmentation of the
testas, or seed-coats, are always associated together on
the same plants ; so that, if we find a plant which has
green leaf axils, we may be sure that its flowers will
be white, and the testas of its seeds only slightly pigmented. On crossing plants bearing coloured axils,
coloured flowers, and pigmented testas, on the one
hand, with plants bearing green axils, white flowers,
and unpigmented seed-coats, on the other, the two sets
of characters are found to behave as a simple pair of
allelomorphs, and the simultaneous appearance of

colour in these different situations doubtless depends upon the presence of a particular pigment in the plant which exhibits it. Nevertheless, we can scarcely fail to look upon these three separate manifestations of the pigment as representing distinct characters, and this being so, we suppose their germinal representatives to be coupled together in such a way that they remain associated at the time when, during the formation of the germ-cells of the heterozygote, other allelomorphs become independently segregated.

And this way of looking at the facts is further justified by the behaviour of the characters in question in another species of plant. For in the sweet pea it is possible for the coupling between these characters to be broken down, so that a plant which exhibits green leaf axils may, under certain circumstances, bear coloured flowers. In such a plant the leaf-axil-colour and the flower-colour must clearly be represented by independent allelomorphs.

In other cases, again, there may be coupling between characters which have no obvious relation to one another at all. In illustration we may take the case of a cross between two strains of peas, one of which had white flowers and opened its buds several days earlier than the second, the blossoms of which were purple.

The F_1 plants (with purple blossoms) came into flower at a period intermediate between those of the parents. In F_2 506 plants were grown successfully. Some of these flowered as early as the white parent, and others as late as the purple parent ; but the majority of the plants ranged between these two extremes, so that it was impossible to rank the individuals into definite classes in respect of so indefinite a character as time of flowering. On making a perfectly arbitrary division, however, it was found that 175 purple and 104 white plants were in flower on a certain day, and that 208 purple and 19 white plants did not open their buds until afterwards. There is, therefore, clearly some coupling between the presence of white blossoms and early flowering on the one hand, and between lateness and purple flowers on the other. Two characters more diverse than colour of the flowers and time of flowering could at first sight scarcely be imagined.

The second class of complications that we have to deal with—although the term complication may be to a certain extent justified in connection with it— does not involve any exception to Mendel's law of segregation. The phenomenon of so-called *reversion on crossing* has long been familiar to biologists. Its meaning, however, was totally obscure, and even the Mendelian was at first unable to offer any explanation. The phenomenon consists in the appearance, in the offspring of a cross, of a character which was not visibly present in either parent, and in many cases this character can properly be regarded as ancestral—it is a character which has been lost by both parents in the course of their divergent evolution from a common primitive form. Now, these cases differ entirely from those of the appearance of a heterozygote form on crossing, such as are due to the combined action of the two parental allelomorphs in the cross-bred offspring, because in true cases of reversion a certain proportion of the reversionary individuals of F_2 are found to breed true, which a simple heterozygote will never do.

It has been found that the essential part of this phenomenon of reversion on crossing consists in the existence in the parents of certain hereditary factors—allelomorphs, in fact—which, although by themselves invisible, yet, when combined in cross-breeding with certain other allelomorphs, belonging to independent pairs, lead to the appearance of new visible characters.

The term reversion cannot properly be applied to these phenomena as a class, because, in the first place, characters may arise in this way which cannot be regarded as ancestral, and, secondly, because reversions may take place in other ways ; for example, the reappearance of a simple recessive character would legitimately be ranked among reversions. The best general name for the class of phenomena we are about to describe is perhaps *latency of characters*, or *cryptomerism*, the latter being the term employed by Tschermak, who was the first to describe these phenomena in connection with Mendelian ratios.

In the simpler cases an invisible or *latent* factor derived from one parent, on becoming associated with a different factor born by the other parent, and already visibly represented among the external features of this second parent, makes itself apparent among the visible characteristics of the heterozygote. In such a case the characteristic appearance exhibited by the heterozygote may subsequently become permanent, owing to the building up of a type which is a homozygote in respect of both the necessary factors.

This may be made clearer by a definite illustration.

A pea-plant characterized by the presence of a greyish or brownish testa to its seeds (grey) was crossed with a plant having nearly colourless testas (white). The testas of the F_1 plants were marked with bright purple dots on a grey ground (purple). These hybrid plants were self-pollinated, and in F_2 the three types appeared in the following proportions : 9 purple, 3 grey, 4 white. What is the meaning of this ratio ? In order to complete the ordinary expectation for a simple Mendelian case in which two pairs of allelomorphs are concerned (di-hybridism) we must write down the following expression :

$$9 \left\{ \begin{array}{l} \text{purple} \\ \text{grey} \end{array} \right\} : 3 \left\{ \begin{array}{l} \text{no purple} \\ \text{grey} \end{array} \right\} : 3 \left\{ \begin{array}{l} \text{purple} \\ \text{no grey} \end{array} \right\} : 1 \left\{ \begin{array}{l} \text{no purple} \\ \text{no grey} \end{array} \right\}$$

But it would seem that the purple character cannot appear when the grey colour, or some factor constantly associated with this colour, is absent, as is the case in the original white parent from which the factor for purple spots was derived. Consequently, the three $\left\{ \begin{array}{l} \text{purple} \\ \text{no grey} \end{array} \right\}$ plants are indistinguishable from the $\left\{ \begin{array}{l} \text{no purple} \\ \text{no grey} \end{array} \right\}$ plants or whites, and we thus arrive at the result which was described as being the one actually obtained—namely, 9 purple : 3 grey : 4 white.

In other respects this example is precisely like the case of two pairs of allelomorphs described on p. 176.

We may write A for presence of grey pigment, a for absence of grey pigment, B for presence of purple, and b for its absence. Then the original cross was of the form $Ab \times aB$, from which $AaBb$ resulted in F_1. And the visible characters of the types which appeared in F_2 would be represented by $9AB + 3Ab + (3aB + 1ab)$. On referring to the account given on p. 176 it will be seen that one in nine of the purple plants is of the constitution $ABAB$, and may be expected to breed true.

A precisely similar result may be obtained in F_2 in cases where there is no reversion in F_1. In the following example a white pea, which did not contain the latent purple factor, was crossed with a 'maple-seeded' pea. The characteristic feature of maple is a marbling of brown spots on a grey ground colour. In F_1 the marbling was dominant, and the seeds resembled the maple parent.

In F_2 there appeared 9 maple : 3 grey : 4 white— *i.e.*, the same ratio as in the previous case, this time without reversion. This ratio is brought about by the simple combinations of two pairs of allelomorphs A-a, and C-c, C being unable to manifest itself unless A is present in the same zygote. As a matter of fact, in this particular case C does sometimes just manage to appear in the absence of A, the result being a white seed with a sort of faint 'ghost' of a maple marking.

When a strain bearing both maple marking and purple spots is crossed with a white in which neither of these factors is latent, we can easily calculate the ratio to be expected in F_2 by using the formula $(A + 2Aa + a)$ $(B + 2Bb + b)$ $(C + 2Cc + c)$. The result works out as follows (writing m for maple, p for purple, and g for grey): $27mpg$, $9mg$, $9pg$, $3g$, ($9mp$, $3m$, $3p$, $1w$). Since g is absent from all the members of the series enclosed in the bracket, these appear white, or nearly so, the total number of whites being thus 16. And the numbers obtained in an actual experiment accorded closely with the expected ratio $27 : 9 : 9 : 3 : 16$.

Among the sixteen whites, some will be bearing the factors for m and p, others that for p only, others that for m only, whilst one in sixteen will contain neither of these factors. Until such invisible differences between the different white plants are actually proved to be present the whole account so far given will remain more or less hypothetical. The proof is obtainable by crossing the different whites with a pure grey strain. The grey factor being thus introduced, the whites which contain a p or an m factor will exhibit the same in their offspring. A number of the whites obtained in F_2 and in later generations were actually crossed with the same grey-seeded plant. Some of the offspring showed both the maple and the purple character, others the maple without the purple, others the purple without the maple, and others, again, sohwed neither; the seeds of these last being exactly like those of the grey parent owing to simple dominance of the grey allelomorph over white.

Another example of the same kind of phenomenon may be taken from the work of the French zoologist Cuénot. In the present instance the experimental crosses were made with animals—namely, mice. As the result of his experiments Cuénot framed the hypothesis that in mice the colours yellow (*J*), grey (*G*), and black (*N*) can only make their appearance when the zygote containing them contains also a perfectly independent colour-producing factor (*C*), which is allelomorphic to, and dominant over, the albino factor (*A*). (According to the notation which we have previously adopted, this last pair of factors would be more properly written C and *c*.)

According to Cuénot's scheme all albino mice (white mice with pink eyes) contain some colour in a latent condition. The different individuals with which experiments were made were found to be capable of representation by the following 'formulæ' of heredity :

Homozygotes, *AG ; AN ; AJ*.
Heterozygotes, *AG . AN ; AG . AJ ; AN . AJ*.

In the visible expression of these colours when *C* is present, yellow was found to be dominant over grey and over black, and grey was dominant over black. In the formation of the gametes the various factors concerned follow Mendel's law precisely. There is, however, one curious point to be noticed, if Cuénot's interpretation is really correct. We have in this case three alternative characters instead of a simple pair of allelomorphs. On referring to the list of different kinds of individuals given above, it will be seen that homozygotes contain only one of these alternative characters, whilst a heterozygote contains two. It is not possible for a single individual to contain all three

13

characters at once. A pair of individuals may, however, between them carry all the possible colour
allelomorphs. Cuénot gives the following example,
being one of three possible methods by which all the
three colours, yellow, grey, and black, as well as the
albino character, can appear among the offspring of a
single pair of animals :

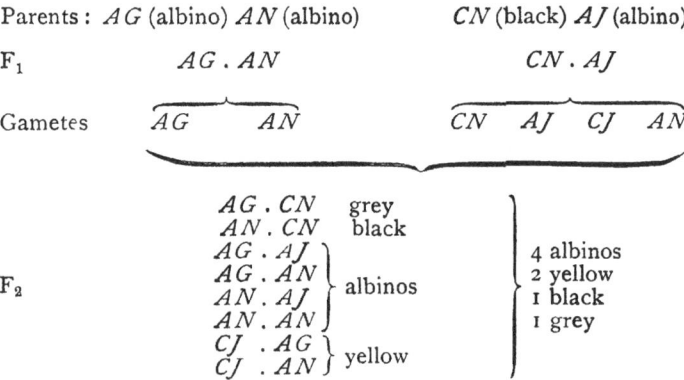

Parents: AG (albino) AN (albino) CN (black) AJ (albino)

F₁ $AG \cdot AN$ $CN \cdot AJ$

Gametes AG AN CN AJ CJ AN

F₂

$AG \cdot CN$ grey
$AN \cdot CN$ black
$AG \cdot AJ$
$AG \cdot AN$
$AN \cdot AJ$ } albinos
$AN \cdot AN$
$CJ \cdot AG$
$CJ \cdot AN$ } yellow

4 albinos
2 yellow
1 black
1 grey

The actual numbers of offspring which were obtained
in F₂ fromt his cross were as follows : 81 albinos,
34 yellow, 20 black, 16 grey ; the expected proportion
being—76 : 38 : 19 : 19.

By the process of crossing followed by selection of
those families in which a uniform series of young
appeared in F₃, Cuénot was able to extract without
difficulty pure homozygous races of blacks $CN \cdot CN$,
and of greys $CG \cdot CG$, as well as the corresponding
albino races $AN \cdot AN$, and $AG \cdot AG$. But in the case
of the yellow a curious and unexpected phenomenon
appeared, and one which, if Cuénot's explanation of it

proves to be well established, is likely to be of great theoretical interest.

When $CJ . CG$ (or $CJ . CN$) is crossed with CJ (or CN), since yellow is dominant, equal numbers of yellow and of grey (or black) offspring are to be expected, and in various crosses of this nature Cuénot actually obtained 177 yellows and 178 blacks or greys. Hence we may deduce that the heterozygote yellow was giving off the expected proportion of gametes bearing the yellow character (*i.e.*, 50 per cent.).

When such heterozygous yellows are bred together the expected result would be as follows :

$$CJ . CG \times CJ . CG = \underbrace{CJ . CJ + 2\ CJ . CG}_{3 \text{ yellow}} + \underset{1 \text{ grey}}{CG . CG}$$

Eighty-one yellow mice were actually obtained in this way. Among them some twenty-seven would naturally be expected to be pure dominant, and to give yellow only when crossed with black or grey individuals. To Cuénot's astonishment, he found on making the necessary crosses that every one of these eighty-one yellows gave some black or grey among its offspring; not one of them was a pure homozygous yellow.

The only way in which this result can be explained at present is by supposing that there is some obstacle to the fertilization of one yellow-bearing gamete by another gamete of the same kind. The combinations $CJ \times CG$ and $CG \times CG$ take place, it would seem, readily enough, but there is some mutual repulsion between CJ and CJ which prevents their union. We shall find later on that there is some evidence derived from an entirely different class of facts which seems to support the idea that a selective fertilization of this kind really does take place in certain cases. The phenomenon is nevertheless so remarkable that we may have some hesitation in accepting it without further evidence. In the meantime it must be recorded as a distinctly exceptional case, though not, be it noted, as an exception to Mendel's law. The gametes obey the law, as was shown by crossing yellow with non-yellow, and it is only in their manner of combination that a complication has been introduced.

We have still to describe a case in which two latent factors, one derived from each parent, give rise, by their simultaneous presence in the zygote produced, to the appearance of an entirely new character. The following example is the first one of the kind to be completely elucidated, and is one of those studied by Messrs. Bateson and Punnett and Miss Saunders.

The white-flowered variety of sweet-pea known as Emily Henderson was found to exist in two forms, only to be distinguished from one another by the shape of the pollen grains which they produced. In one of the two the shape of the pollen is elliptical (long pollen), in the other it is approximately spherical (round pollen). Sweet-peas normally undergo self-pollination, so that the two types naturally remain distinct. Let us see what happened when the long- and the round-pollined forms were crossed together.

The cross-bred plants (F_1) had coloured flowers— flowers of the old-fashioned purple type known to florists as Purple Invincible, which is characterized by a purple standard and blue wings. The pollen produced by these plants was of the long type. Thus as regards the shape of the pollen grains there was simple dominance. But the union of two white-flowered types has given rise to a series of plants all possessing a definite colour character—purple with blue wings. This character is very probably the same as that exhibited by the common ancestor of all our cultivated sweet-peas. Here, then, we seem to have a clear case of reversion to the ancestral type on crossing. We shall find that the Mendelian principles will enable us to arrive at a clear conception of the mechanism of this process.

The cross-bred plants were self-pollinated, and in F_2 the following types made their appearance in approximately the proportions given :

Purple Invincible	...	81	or	3	3		
Picotee	... ...	27		1		9	
Painted Lady	...	27	or	3	1		
Tinged white	...	9		1			7
White...	... ...	112					

Painted Lady is a well-known colour type which is characterized by the presence of a red standard and white wings. Picotee and tinged white are also forms well known to the sweet-pea fancy. They appear to be diluted forms of the purple and Painted Lady types respectively, their appearance depending upon the presence of a definite diluting factor in addition to the factor for the colour in question, or perhaps more properly upon the absence of the proper strengthening factor which converts Picotee into purple, and tinged white into Painted Lady.

The following explanation of the result so far described has now been well established by further experiment. In the first place, we may consider all the coloured forms together as a single group opposed to white. It is now clear that the coloured type of F_1 is due to the meeting together of two factors, one of them born by one white parent and the other by the second, and it is necessary for both these factors to be simultaneously present in order that any colour may make its appearance. We may call these two factors C and R, denoting the absence of either by c and r respectively. By the simple Mendelian behaviour of these two pairs of factors C-c and R-r, the ratio of nine coloured plants to seven white appearing in F_1 is readily explicable, and the way in which this happens is shown in the diagram on the opposite page.

To explain the presence of the four different types of coloured plants which make their appearance in F_2, two further pairs of allelomorphs are called in. The dominant member (B) of one of these, when present in combination with C and R, produces the purple or Picotee colour (blue), whilst its absence (b) in presence of C and R is accompanied by the appearance of the red colours—Painted Lady and tinged white.

Purple Invincible and Painted Lady are regarded as intensified forms of Picotee and tinged white respectively. The presence of the second factor (T) is attended by the development of the full colours purple and Painted Lady ; its absence (*t*) causes the appearance of the diluted forms Picotee and tinged white.

B and *T* may be present when either *C* or *R* or both are absent ; the resulting plant has then white flowers. And it is interesting to notice that the ultimate recessive white, containing *c r b t*, occurs only once among 266 individuals of F_2. The whole apparently complex system

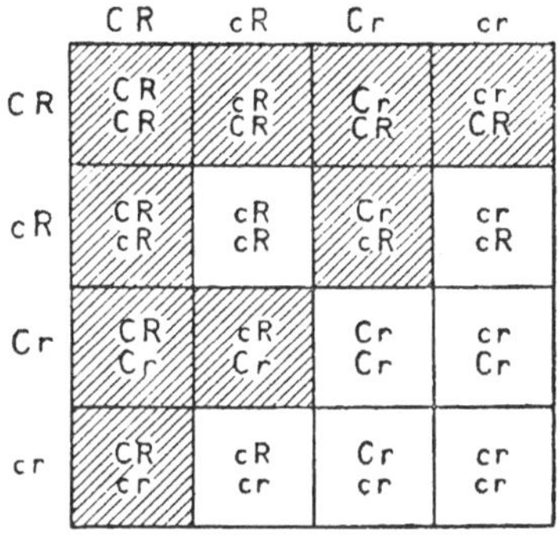

	C R	c R	C r	c r
C R	C R / C R	c R / c R	C r / C R	c r / C R
c R	C R / c R	c R / c R	C r / c R	c r / c R
C r	C R / C r	c R / C r	C r / C r	c r / C r
c r	C R / c r	c R / c r	C r / c r	c r / c r

FIG 16.

The shaded squares represent coloured plants, the blank squares white plants.

of floral colours is thus explained by the simple Mendelian behaviour of four separate pairs of allelomorphs.

Bateson and his collaborators have, therefore, provided a complete account of the phenomenon of reversion on crossing, an account which has already been demonstrated to hold good in other instances besides that of the sweet-pea. The facts are expressed in the following manner by their discoverers. ' " Reversion " is thus seen to be a simple and orderly phenomenon due to the meeting of factors belonging to distinct though complementary allelomorphic pairs, which at some moment in the phylogeny of the varieties have each lost their complement.'*

We may now proceed to pass in rapid review a selection of the more remarkable instances of Mendelian inheritance which have been so far demonstrated.

The ease with which characteristics of colour can be distinguished and defined has naturally led to a good deal of attention being paid to the phenomena of their inheritance. In this way many cases of simple dominance have been discovered in plants and in animals, as well as several examples of reversion in F_1, followed in both cases by a Mendelian segregation of characters.

Thus the colours of many flowers afford perfectly simple phenomena, whilst other cases, like the sweet-peas and the closely similar case of stocks studied by Miss Saunders, have required long and arduous experiment for their elucidation. No case of this kind hitherto examined has been definitely proved to be non-Mendelian.

Colour characters which follow Mendel's law have been observed in mice, rats, rabbits, guinea-pigs, pigeons, fowls, cats, and so on. In butterflies and other insects, and even in snails, similar phenomena have been descried. The study of the larger domestic animals awaits for the present the proper endowment of these researches. When this takes place the inheritance of far more important characters than colour will be adequately studied to the great profit of all who are concerned in the breeding industry.

* Proceedings of the Royal Society, B. vol. 77, p. 238.

Hurst has already shown from an examination of the stud book that the bay and brown colours of thoroughbred horses are Mendelian dominants to chestnut.

Other characters of the most diverse kinds are also similarly inherited. We have already referred to structural characters in maize and in peas. Stature is a character which is definitely inherited in many plants. Among more subtle characters a similar mode of transmission has been found in the case of differences in chemical composition, and in that of immunity from and susceptibility to the attacks of certain diseases. The thrum-eyed condition of the primrose has been shown by Bateson and Gregory to be a Mendelian dominant to the pin-eyed condition, so that we have here the solution, so far as solution is possible, of a biological problem to which Darwin devoted the greater part of a volume.

A study of numerous pedigrees has enabled Bateson to show that there is great probability that in the case of the human race certain congenital diseases are simply transmitted from parent to offspring in accordance with Mendel's law.

How far the influence of the Mendelian principles may extend we do not yet know. But it is certain that very few, if any, cases have so far been discovered in which differentiating characters do not behave in this way when the types which exhibit them are crossed together. Experiments have now been made upon a great variety of plants and animals, involving a considerable diversity of kinds of characters. Nevertheless it is scarcely possible to cite a case in which it is definitely and certainly known that Mendel's law, subject to the modifications already described, does not hold good. Cases of various kinds are, indeed, recorded, but these records are derived from experiments either carried out before the bearings of the Mendelian phenomena were at all fully appreciated, or —and this is the most frequent case—without any knowledge at all of Mendel's discovery.

Thus a considerable number of cases were formerly described in which the first cross or heterozygote of F_1 bred true instead of segregating in F_2. There is some doubt whether any case of this kind will really stand criticism; Millardet's case, for example, which was described at the end of the last chapter but one, has never been confirmed. It is quite certain that among all the numerous crosses studied during the last six years no example of the kind has been substantiated. The most recent cases to be described of a first cross breeding true are those of de Vries, and at these we are bound to pause, because de Vries is surpassed by no recent observer in weight of authority. Nevertheless, de Vries' cases are of so complex a kind that we have some hesitation in accepting them without further study. For the rest, this is one of the problems which remain for the future to deal with.

We may now turn for a brief space to some of the cases in which we have as yet no certain knowledge of the manner in which inheritance proceeds.

The most obvious extension of Mendel's law to processes where it cannot be directly shown to hold good is to suppose that the same rule applies to cases of normal fertilization as to hybrid fertilizations. We should then picture the former process as taking place in somewhat the following way. Every visible character of the individual which can be separately distinguished, and which on cross-breeding would be inherited on ordinary Mendelian lines, must be represented in the gametes by a definite factor of some kind, possibly by a definite substance or combination of substances. The pair of parental factors for a particular character would combine on fertilization, and at the formation of the gametes in the offspring its members would separate as perfectly definite entities, to recombine when these gametes meet once more with their corresponding mates. Such a definite segregation of characters taking place within a pure strain would be very difficult of absolute demonstration, but it is hard to avoid the conclusion that this is a true deduction from the facts observed when cross-breeding takes place. Such a segregation would formerly have been thought a very small assumption in comparison with that of the segregation of pairs of allelomorphs of which no trace is externally visible, and yet the latter assumption has now been shown to be perfectly well established.

This idea of unit characters, capable of being inherited independently of one another, is one of the most important conceptions which has ever been introduced into the science of biology, and the introduction of it has followed as the direct result of Mendel's work. It is a conception which has led to a complete change in our ideas of heredity, since we no longer look upon the individual as a unit, but find ourselves compelled to study separately the independent characters of which the individual is built up. The idea of the individual as a living mosaic—an idea put forward long ago by Naudin with only a partial realization of its significance—has thus returned to us. In this connection a curious problem presents itself. What would be left if we could imagine all the separable characters of a living creature as having been taken away ? Would there, or would there not, be any residuum ? Upon this knotty point there is a disagreement among authorities, and so we may be content to leave it, since the question is hardly one which is capable of a practical solution.

A phenomenon to which it is scarcely doubtful that Mendelian principles will ultimately be found to apply, although as yet the precise proof is wanting, is that of sex. In the male and female sexes of the majority of animals we have a very clear example of a pair of definite differentiating characters. And the fact that in the majority of forms the two sexes make their appearance in nearly equal numbers, may be thought to point clearly to the conclusion that the separation of the sexes depends upon some quite simple gametic process. Light has recently been thrown upon this question from the side of the study of the minute structure of the gametes, and we shall defer the further discussion of the problem to the chapter which deals with microscopic phenomena within the cell.

A proper understanding of Mendel's law enables us

to escape certain theoretical difficulties which have long been prominent in the minds of students of evolution. Many evolutionists were accustomed to argue that a new form suddenly arising in the midst of an old-established species could not give rise to a new and permanent variety or elementary species, because it would immediately be 'swamped' by intercrossing with the parent species from which it was derived. If, however, the character distinguishing the new type is allelomorphic to the corresponding character, or absence of a character, shown by the parent form, this difficulty disappears. For suppose as an extreme case that the new type arises as a single individual only, which is therefore compelled to mate with a member of the original species. If the new character is recessive it will disappear in the immediate offspring of this cross. But half the germ-cells produced by the cross-bred form will bear the new character pure and undiluted. If any of these cross-breds mate together the new type will appear in a quarter of their offspring. Even if all of them mate with members of the original type, half the offspring of such matings will be heterozygous, and sooner or later the heterozygotes will be sure to mate with one another, and give rise once more to the novel type of individuals. If the new form has any structural or other advantage over the old species, the former will tend to survive at the expense of the parent type, and it may survive if it is only equally well fitted for the battle of life. In the case of dominance of the new form the same process will take place, only it will be apparently more rapid in the early stages because the cross-breds will themselves exhibit the new character. In this case, even if the new type has a very marked advantage over the parent form, the process of completely supplanting the latter will be delayed, because the old type of character can survive concealed in heterozygote individuals.

Let us pause for a moment to sum up the novel ideas which have so far been presented in this and the preceding chapter.

We found in the first place that from the point of view of heredity we must look upon an animal or a plant as a composite being, made up of a great number of unit characters, each capable of separate description, and all inherited independently of one another

When a pair of nearly-related animals or plants are mated together, when, in fact, like is bred with like, and with still greater certainty in cases of self-fertilization such as are not uncommon among plants, every unit character born by one gamete finds a corresponding mate among the characters born by the second gamete. It naturally follows that a series of characters similar to those of the parent or parents make their appearance in the offspring.

When a pair of individuals belonging to distinct varieties or races are mated together, the result is the same in the case of the majority of characters exhibited by each of them. For separate varieties of the same species differ from one another in a small number of units only, and organisms which differ in more than a few unit characters refuse altogether to unite for the production of offspring. From the study of the precise behaviour of those characters in which a pair of parental organisms differ, a flood of light has been thrown upon the phenomena of inheritance.

We find, as a rule, that opposed to every differentiating unit character of one parent there exists a corresponding but different character in the other parent. One parent may have smooth seeds and the other wrinkled seeds, for example. Very frequently the corresponding feature consists in the absence—or failure to appear—of a particular character, as, for instance, when the non-development of pigment leads to the appearance of white flowers.

We can now realize how necessary it is, in order to avoid hopeless confusion, to follow the behaviour of each pair of characters in the offspring separately.

The result of the meeting between the two opposed characters of the same pair we saw to be different in different cases. There may arise in the offspring (1) the appearance of a simple blend of the two parental characters. Or (2) one character may be more or less dominant over the other. Or (3) the combination of the two parental characters in the offspring may give rise to an appearance quite different from that of either of them, very much in the same way as in chemistry oxygen and hydrogen when combined give rise to water. Or (4) we may get further complications in which unsuspected characters, present in an invisible condition in one or both parents, take a part, often giving rise to the appearance of a supposed reversion.

The most important phenomenon of all, however, is that which is found to occur at the formation of the germ cells of the heterozgyote plant or animal. Whatever the appearance of the hybrid form may have been, at this stage in its history the determining factors for each member of the pair of parental allelomorphs reappear in their entirety in certain cells which by their diversion give rise to the gametes, and at one of the divisions in question the parental characters (in a potential condition) separate completely from one another, so that half the gametes bear one allelomorph and half of them the other. In cases where more than one pair of allelomorphs has taken part in the cross, the members of each pair are found, as a rule, to undergo this process of segregation quite independently of all the other pairs.

Whether this process of Mendelian segregation is a universal one, we cannot tell at present, but we do know that it is of very widespread occurrence. Nor do we know whether a similar process takes place at the gamete-formation of homozygotes, though it seems scarcely possible to suppose otherwise.

We have seen enough to enable us to recognise very clearly the vital importance of an understanding of the constitution of the gametes in all questions of heredity. There must exist in the gametes, in an uncombined condition, those units which by their combination in zygotic organisms lead to the appearance of the characters which we can recognise. But we have seen that, owing to the appearance of dominance and other kindred phenomena, the visible external characters of an organism are not a complete guide to the nature of its gametes. It is only by careful breeding that we can distinguish the heterozygote from the pure dominant foim—to take the simplest possible example of this difficulty. For this reason it has now become the chief business of the student of heredity to determine by experiment what combinations of allelomorphs are present in the gametes of the individuals with which he is working.

The behaviour of these allelomorphs has now been disentangled in many cases of very considerable complexity ; and all such cases as have been so far examined in detail have proved explicable in terms of a larger or smaller number of allelomorphic pairs, all of which obey Mendel's law—with the single exception of those cases in which coupling between the allelomorphs of different pairs introduces a slight further complication. Although it is perhaps scarcely probable that Mendel's law will ultimately prove universal in its application, nevertheless the few exceptions recorded by competent observers still require further examination before they can be accepted as invalidating the law in any single instance.

The question naturally arises as to how far the Mendelian rule of inheritance agrees with or contradicts those estimations of hereditary values which have been arrived at by the labours of the biometricians.

So long ago as 1902 Mr. G. Udney Yule endeavoured with some apparent success to reconcile the Mendelian results with those of biometry. Progress has been rapid during the last four years, and what we have now before us is rather the question of reconciling the biometrical conclusions with the firmly-established facts of Mendelian inheritance. Quite recently Mr. Yule seems to have succeeded in performing this service for science, although the comments of other biometrical students upon his work have still to be awaited.

In 1902 Yule considered the case of a pair of simple Mendelian characters, A and a, exhibited in a mixed population breeding together at random, in such a way that the total number of germ cells bearing A and a respectively might be regarded as equal in any generation. In such a case it will always be an even chance whether a recessive parent will produce a dominant or a recessive child, because the chance of its gamete (a) mating with A or a is the same. A knowledge of the ancestry of the recessive parent makes no difference to the result. Consequently the case of the pure recessive does not fall in with any possible theory of ancestral heredity.

But on turning to the dominant parent, the case is found to be different. For such an one may be either a pure dominant homozygote giving off A-gametes only, or it may be a heterozygote giving off equal numbers of A- and a-gametes. Yule shows that if both the parents of the A individual exhibited the character A, the proportionate number of its offspring which may on the average be expected to show the A character is greater than would have been the case if one of its parents exhibited the character a. And in a similar

way a knowledge of the characters shown by the grandparents adds something to the certainty of the pre-
diction as to the proportionate numbers of offspring of
the two kinds which are to be expected, when the
average of a number of cases is taken according to the
usual statistical method.

Yule therefore regarded the case of the dominant
character as showing conformity with the law of
ancestral heredity, according to his own statement of
that generalization, which was to the following effect :
The law that ' *the mean character of the offspring can be
calculated with the more exactness, the more extensive
our knowledge of the corresponding characters of the
ancestry*, may be termed the law of ancestral heredity.'*

It may be remarked in passing that Yule's distinction of the problems of genetics into those of
intra-racial heredity and those of hybridization cannot
now be regarded as holding good, unless the term
hybridization is to be extended to many cases—*e.g.*,
that of the inheritance of coat colour in thoroughbred
horses, which would have been classed unhesitatingly
as instances of heredity by all biometricians in 1902.
Bateson's instinct did not fail him when he divided
these problems into those of continuity and those of
discontinuity respectively, although at the present time
the realm of continuous variation and inheritance is
being steadily encroached upon owing to the analysis
of complex characters into definite constituents.

In 1904 Karl Pearson struck a blow at the prospect
of conformity between biometrical and Mendelian

* ' New Phytologist,' vol. i., p. 202.

results in his memoir, 'On a Generalized Theory of Alternative Inheritance, with special reference to Mendel's Laws.' Pearson's treatment of the subject involved advanced mathematical reasoning, and we can, therefore, only give a brief summary of his main results. Pearson proposes special terms for the *A* and the *a* elements respectively of a couplet or pair of allelomorphs. He proposes to call the *A* element a *protogene*, and the *a* element an *allogene*, and he thus distinguishes between the two sorts of homozygotes by calling *AA* a *protozygote* and *aa* an *allozygote*.

Pearson considered the case of a population breeding together at random, in which a single measurable character, such as stature, is determined by the combined action of an indefinite number of pairs of allelomorphs, and he proceeded to work out the value of parental correlation which was to be expected under these circumstances. This value he found to be exactly one-third, a value which happens to be identical with Galton's original determination of parental correlation from his statistics of human stature. A considerable number of determinations of parental correlation have, however, since been made in the case of all kinds of characters. The values show considerable variation, but the average which they indicate is much nearer to 0·5 than to 0·33. Pearson therefore concluded that in none of these cases could anything resembling Mendelian inheritance be taking place, and that the latter is, in fact, the exception rather than the rule.

Mendelians, aware of the certainty of their own

results, and being convinced that these facts must have a very wide application, were thereupon driven reluctantly to the conclusion that something was seriously wrong with the methods adopted by biometricians for determining the coefficients of correlation. It seems, however, that this conclusion may have been arrived at with undue haste.

In August of the present year (1906) Mr. Yule read a very interesting paper before the International Congress of Hybridization assembled in London on 'The Theory of Inheritance of Quantitative Compound Characters on the basis of Mendel's Laws.' Though some difficulty was then experienced in following his argument by an audience unaccustomed to statistical methods, Yule's conclusion is really very simple.

Yule points out that the only character dealt with in Pearson's memoir is the number of protogenic or allogenic couplets present in the individual, and it is the proportionate number of these couplets present in the parent and in the offspring respectively which is taken as determining the value of the correlation coefficient. Consequently Pearson's treatment of the subject does not justify his statement that the Mendelian theory gives a rigid value for the coefficients of parental correlation for all races and characters—a conclusion which he regards as fatal to this theory, because the coefficients for different characters and races, as found statistically, show considerable individual differences, and seem to cluster round a value considerably higher than that indicated by his elaboration of the theory of

the pure gamete. Yule thereupon discusses a somewhat
more general case, and considers the inheritance of a
length made up of a number of distinct segments, each
of which is determined by an independent pair of
allelomorphs. Supposing each segment to take the
length a, b, or c, according as the corresponding protozygote, heretozygote, or allozygote is present, Yule
arrives at an equation from which the correlation
between parent and offspring may be found. From
that equation the following results are deducible :

If there is dominance—$i.e.$, if $a = b$, or $b = c$, the correlation coefficient is the same as that found by Pearson
—$i.e.$, one-third.

But if the heterozygote always gives rise to a
length exactly intermediate between those due to
the respective homozygotes, the correlation is found
to be one-half.

Cases of partial dominance will give an intermediate
value. Consequently, according to the degree of
imperfection of dominance, and without assuming any
other disturbing circumstances, values of parental
correlation varying from 0·33 to 0·5 are to be expected
on the Mendelian theory of inheritance when applied
to populations. These figures are calculated on the
supposition that there is random mating of the parents,
but if there were a tendency for like to mate with like
the correlation values would become still higher. Yule
therefore concludes that ' there is therefore no difficulty in accounting for a coefficient of 0·5 on the
theory of segregation, but such a value probably
indicates an absence of the somatic phenomenon of

dominance. In the case of characters like stature, span, etc., in man this does not seem very improbable.'

It is impossible to bring the present chapter to a conclusion without some reference to the practical aspects of the Mendelian discovery. The progress of experimental research in this field during the last half-dozen years has been so rapid, that there is little ground for astonishment in the fact that only a small proportion of those to whom the discovery of the Mendelian method is of the very highest importance from a commercial point of view have yet arrived at any serious appreciation of it. The improvement of the breeds of cultivated plants and domestic animals is a subject of vital importance to the whole human race, quite apart from the question of the commercial profit which it represents for those whose business it is to be directly concerned with the process—the actual plant- and animal-breeders themselves.

Hitherto the methods of amelioration which have been adopted have depended largely upon guess-work, or at the best upon the result of practical experience. We are now within sight of the day when a complete system of precise scientific methods will have been elaborated. The time required for the development and application of these methods must chiefly depend upon the apathy or enterprise of those in whose hands rests the means of subsidizing this kind of work, for without proper resources the progress of any such study must of necessity be slower than it would be if properly-equipped establishments were at the disposal of duely-trained experimenters receiving an adequate remuneration.

The practical application of Mendelism cannot be better illustrated than by an account of Mr. R. H. Biffen's work upon the improvement of cereals, particularly of wheat—work which exhibits an extraordinary contrast in point of scientific exactness with everything of the kind which has been previously undertaken. This contrast was remarkably displayed at one of the morning sessions of the recent International Congress on Hybridization and Plant Breeding, held under the auspices of the Royal Horticultuial Society. On that occasion a series of communications upon the subject of cereals culminated in an admirable account given by Mr. Biffen of the way in which the problems of their improvement have been overcome at the experimental farm of the Cambridge University Department of Agriculture. And it was a gratifying sign of better times to observe the enthusiastic interest with which practical men greeted his communication.

As a preliminary measure Biffen has worked out the inheritance of a number of comparatively simple characters, many of which have little practical importance. But the fact of their strictly Mendelian behaviour showed the possibility of readily obtaining any desired combination of them, and at the same time rendered it highly probable that characters of a more practical value to the farmer would prove similarly amenable to the breeder's art.

Thus Biffen found that the following pairs of

characters, among others, exhibited simple Mendelian phenomena, the one placed first being in each case the dominant :

Beardless ears.	Bearded ears.
Keeled glumes.	Round glumes.
Felted glumes.	Glabrous glumes.
Red chaff.	White chaff.
Red grain.	White grain.
Thick and hollow stem.	Thin and solid stem.

And so on. In other cases, again, the F_1 generation showed a character intermediate between those of the parents, and in F_2 there appeared a ratio corresponding to $A : 2Aa : a$.

Thus when Polish wheat (early) was crossed with Rivet wheat (late), the time of ripening of the F^1 generation was intermediate between those of the parents. In F_2, 103 early, 210 intermediate, and 100 late plants, were counted. Time of ripening is, moreover, clearly a character which may be of considerable practical importance.

In further illustration of what can be done from a commercial point of view, we will consider the case of two other characters only—rust immunity and ' strength.'

There is a quality of wheat grains known as *strength* which is essential for the production of a flour such as can be baked into the kind of loaf which is at present the only one saleable in England. This quality unfortunately happens to be wanting in all the strains of wheat which it has hitherto been possible to grow at a profit in this country. For this reason imported American and Canadian hard wheats, which contain this quality of strength, are worth in England some shillings a quarter more than home-grown wheats.

When such strong American varieties are grown in this country the majority of them are rapidly found to lose this quality, and to become after a short time as 'weak' as ordinary English wheats. Some of them do, however, retain their strength, and after several seasons —in one case fourteen—show no signs of deterioration. An example of a wheat of this latter type is afforded by Red Fife, which is the basis of the mixed wheat known commercially as Manitoba Hard, the latter consisting, as a matter of fact, of a mixture of several different varieties. Unfortunately these permanently hard wheats do not yield so large a crop as the commonly cultivated English varieties, and so their higher price does not make up for the smaller number of bushels per acre obtained when they are grown.

Biffen therefore set to work upon the problem of combining hardness or strength with the power of yielding a good crop, and with the other desirable qualities characteristic of the home-grown varieties. With this end in view Manitoba Hard was crossed with a typical English wheat—Rough Chaff.

The F_1 plants produced grains all of which were fully as hard as those of the Manitoba variety.

These grains were sown, and it was found that some of the resulting plants produced strong grains and others weak ones, and that the former were to the latter very nearly in the numerical ratio of 3 : 1. Actually they were as 152 : 48 in a sample of 200 taken at random.

In order to obtain confirmation of this most important result, Mr. Biffen sent samples of the grains born by the F_2 plants to a well-known authority on milling wheats, requesting his judgment upon them, but without telling him their manner of origin. The answer was even more satisfactory than could possibly have been anticipated. Certain of the samples were stated by the expert to belong to the variety Red Fife, which is the name of the particular strain of Manitoba Hard originally made use of in the experiments, whilst others were assigned to a definite strain of ordinary weak English wheat. The segregation of these characters was, therefore, complete, strength being a Mendelian dominant to weakness.

In the next generation certain of the dominant plants, as was to be expected, bred true, and amongst them were individuals which combined with strength of grain the other desirable qualities of the second parent. The problem has, therefore, been completely solved, and there can be little doubt that when these new types are brought into general cultivation the profit obtainable from the growing of wheat in this country will be increased by several shillings to the acre of crop grown.

We may next turn to an even more important achievement. In many countries the annual loss of crop due to the attacks of yellow rust, *Puccinia glumarum*, amounts on a moderate estimate to a considerable number of millions of pounds sterling. Certain strains of wheat exist, indeed, which are more or less completely immune to the ravages of this fungus, but these are usually wanting in other qualities which are indispensable to the farmer. If it should be found that immunity to rust is a simple Mendelian allelomorph, it would be possible to combine this quality with any other useful character which obeyed the same law of inheritance—as several useful characters have already been shown to do. At one time it must have been thought that a similar method of inheritance of the character rust-immunity was too excellent a boon to be reasonably hoped for.

Among a great number of strains of wheat grown on the Cambridge experimental farm, several types showed marked differences in the degree of their immunity from, or susceptibility to, the attacks of *Puccinia glumarum.* Among them Mr. Biffen found one which was apparently quite immune, and, though grown in the midst of numbers of rusted plants, itself never showed a trace of infection. Of another type, known as Michigan bronze, no single individual ever escaped the rust, and so badly were the plants of this strain diseased that very few ripe grains could ever be obtained from them.

Biffen crossed these two types together. In the first generation every plant without exception was badly rusted, but fortunately a considerable number of ripe grains was obtained, and these were sown to produce the second generation. When the plants of this generation had grown up it was observed that among a majority of badly-rusted plants certain individuals stood out fresh and green, being entirely free from infection. On examination it was found that every plant could be placed in one or other of two categories—either it was badly rusted or it was entirely free from rust ; and the numbers of the two kinds of plants were as follows : 1,609 infected, 523 immune.

It is clear, then, that immunity and susceptibility to the attacks of yellow rust behave as a simple pair of Mendelian characters, immunity being recessive. And it is, therefore, possible to obtain by crossing, in three generations, a pure rust-free strain containing any other desired quality which is similarly capable of definite inheritance.

CHAPTER IX

EVERY living creature may be regarded as being built up of a number of structural units which are known as cells. In the case of some of the simplest animals and plants, indeed, the whole body of the organism is composed of a single cell—a small mass of living protoplasm, containing, as a rule, only one nucleus. But in all the higher animals and plants the adult body is made up of a great number of such cells living in intimate association with one another.

The living material of which the cell is composed is known as *protoplasm*. Protoplasm is a highly complicated and unstaple combination of substances, amongst the constituents of which the chemical elements, carbon, oxygen, hydrogen, nitrogen, and sulphur, play the chief parts. Its consistency is slimy and semifluid.

Concerning the nucleus—the most essential and characteristic of cell organs—more will have to be said later on. Other important organs of cells are a wall or membrane which externally surrounds them, one or more vacuoles or cavities containing a watery fluid, or sometimes a gas, and a certain number of more solid bodies or *plastids*. Certain plastids present in the majority of plants are of particular importance as containing the green substance chlorophyll, which plays an essential part in the fixation of carbon from the atmosphere.

Amongst unicellular organisms — the creatures already mentioned as being made up of a single cell only—those which contain chlorophyll and are provided with a firm cell wall, built up of a material known as cellulose, are usually regarded as simple plants; whilst those in which chlorophyll and a cell wall are absent are looked upon as simple types of animals. Similarly slight differences distinguish the cells which build up the fabric of the higher plants from those of which the bodies of the more complicated animals are composed, so that in almost all essential points an account of the behaviour of the cells of the members of one kingdom will apply equally well to those of the other. After a few further preliminary remarks we shall, therefore, for the sake of simplicity, speak of a generalized type of cell, the behaviour of which, except in points of detail, will resemble that of the actual cells of plants or animals indifferently. But in order to convey a more definite idea of an unicellular animal to those who are unfamiliar with the now flourishing science of Protozoology, we may refer briefly to the well-known form *Amœba*, which will serve as an excellent type of an animal consisting of a single free-living cell.

This little creature consists of a mass of protoplasm enclosing a nucleus which is more or less centrally situated and approximately spherical in form. The protoplasm is divided into an outer hyaline and an inner granular portion, the former being limited externally by a very delicate membrane. The shape of the animal is irregular, and, moreover, undergoes gradual alteration owing to the characteristic amœboid movements. These consist in a slow protrusion and withdrawal of processes of the body, enabling the animal to change its position by a kind of flowing movement, and also to engulf its food, which consists of various

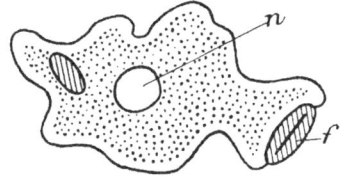

FIG. 17.—AMŒBA.

n, Nucleus; *f*, food particle.

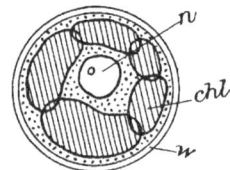

FIG. 18.—PLEUROCOCCUS.

n, Nucleus; *w*, cell-wall; *chl*, chloroplast.

minute organic particles, by the simple process of flowing around it.

In contrast with *Amœba* the unicellular plant *Pleurococcus* is motionless, and is surrounded by a firm wall of cellulose. In addition to a central nucleus, the plant contains, embedded in its peripheral protoplasm, several plastids which bear the chlorophyll concerned in the assimilation of carbon from the gases of the atmosphere. This chlorophyll lends a green colour to the whole contents of the cell, and in its natural habitat the plant is quite conspicuous. The green powdery substance often to be seen on the bark of trees, especially on the side turned to the north, and in similar shady situations, consists, as a rule, of great numbers of minute *Pleurococcus* plants, although the size of a single specimen may be represented by a diameter of little more than the two-thousandth part of an inch.

We are more particularly concerned, however, with the higher animals and plants, the bodies of which are built up of a great number of separate cells. Some of these cells may be modified in various ways, but they all conform, at least in the youthful condition, to types not far removed from those of *Amœba* and *Pleurococcus* respectively. Certain parts of these higher organisms, indeed, such as the bones of vertebrate animals and the wood of trees, do not consist solely of living cells, but are composed to a great extent of dead material excreted or built up by the activity of living cells. These latter have, then, either ceased to live, or they may continue to exist in the interstices of the hard skeletal framework.

New cells come into existence in only one way— namely, by a process of division which takes place in a pre-existing cell. In comparatively rare cases a cell may give off a small bud which forthwith develops into a new cell like the old one. In such a case we may speak of the cell which gives off the bud as the mother- cell, and of the cell into which the bud develops as the daughter-cell. But by far the most frequent method of cell-reproduction, and the only one which is characteristic of the higher animals and plants, takes place by the equal division of an old cell into two

15

new ones. In this case, it is only by a stretch of language that we can speak of parent- and daughter-cells, for the individuality of the pre-existing cell is completely lost, and two fresh individualities have now taken its place.

Since all the cells of the animal or plant body arise by the bipartition of pre-existing cells, it is clear that if we follow these processes far enough back, in the case of any individual organism, we may arrive at a period at which only one cell was present. And under ordinary circumstances this is actually the case. Every individual among the higher animals and plants, arising by the ordinary sexual method, existed at the earliest stage of its individual history in the form of a single cell, the fertilized ovum. And the first obvious process in the development or embryology of the young organism consisted in the division of this embryonic cell into two new cells. Each of these new cells then divided again in like manner, and the multiplication of cells continued until all the innumerable cells which build up the organs of the adult body had finally come into existence. When growth is completed cell-divisions continue more slowly, producing new cells to make good the wear and tear of the bodily tissues.

As the number of cells increased, their relation to one another in space was constantly changing. Different cells, too, became modified in different ways; for instance, the cells on the outside of the young embryo took on a different form from those within, in accordance with the different conditions to which they were exposed, and a host of other changes took place too numerous for us to follow in detail. Thus the complicated structure of the adult organism was gradually arrived at by a process of development in which cell-multiplication played a most prominent and essential part.

We have next to inquire what is the method of origin of the original embryonic cell—the fertilized ovum—from which the new animal or plant develops.

As is indeed implied by the expression 'fertilized ovum,' this cell arises by the fusion together of two independent cells, such fusion constituting the process of fertilization or impregnation. One of the cells which took part in the fusion was derived from one parent organism, and bore the distinguishing characteristics of the cells which composed that parent—or at least some part of those characteristics—whilst the other was in like manner derived from the second parent.

It is to be observed that this fusion together of a pair of cells, derived (in the case we are considering—namely, that of ordinary biparental reproduction) from two separate individuals, results in the formation of a complete new individuality, which arises definitely at that point of time at which the fusion of the two conjugating cells takes place. In this way the cells of the offspring are seen to be of double origin, and it is found that traits and characters derived from both the father and the mother can co-exist in them side by side.

The cells which take part in the above-mentioned fusion are known as *gametes*, or *germ-cells*—male and

15—2

female respectively, according to the sex of the parent from which each is derived. In animals the female gamete is known as the *ovum*, and the male as the *spermatazoon*, and the product of their fusion, as already said, is called the fertilized ovum. Germ-cells of a similaɪ kind arise in a slightly different way in plants. The germ-cells are produced in special parts of the organism known as the generative organs, which in flowering plants are represented by the pistils and stamens.

A more convenient expression for the fertilized ovum is that of *zygote*, a term which we have previously encountered in the shape of the homo- and hetero-zygotes of the Mendelian. By an expansion of meaning the term zygote is also used to express the whole organism which ultimately arises from the product of fusion of a pair of gametes, and by this use the importance of the gamete, as opposed to the zygotic organism as a whole, is brought into due prominence.

We find, then, that the succession of generations in the higher animals and plants, according to the common use of this expression, depends upon the succession of a much larger number of cell-generations. By repeated divisions, each giving rise to a new generation of cells, the fertilized ovum gradually develops into the adult organism. By the division of certain members of the later generations of cells which compose this organism the gametes are produced. By the conjugation of a pair of gametes a zygote of the second generation arises, and the same processes are continually repeated.

Each of the cells hitherto referred to possesses a single nucleus, which is usually a more or less spherical body occupying a central position within the cell. Nuclei, like the cells which contain them, arise only by the division of pre-existing nuclei. Thus the history of the nuclei is in every way similar to the history of the cells, of which they constitute so important a part. In fertilization the nuclei of the conjugating cells or gametes fuse together to form the single nucleus of the fertilized ovum, and every division of this cell, as well as of its cell-progeny, is preceded by a division of the nucleus into two similar portions.

We may forthwith concentrate our attention upon the nucleus as being that part of the cell which is of primary importance from the point of view of heredity, for it is now generally recognised that the nucleus is the part of the cell in which hereditary features are in some way carried. And we may next consider a little more closely the structure of the nucleus as seen under high powers of the microscope.

In what is somewhat improperly called its resting condition—a condition which is characteristic of nuclei at all times when they are not actually undergoing division, or preparing for that process—the nucleus may be seen to be bounded by a more or less definite *nuclear membrane*. The internal structure of such a nucleus is described as reticular—that is to say, at least two different substances are differentiated within the nucleus, one of them forming a reticulated meshwork, the interspaces of which are occupied by the other (Fig. 19).

In entering into a detailed description of the changes which take place in the finer structure of the nucleus, it must be clearly understood that the more minute features alluded to are only to be seen with any degree of definiteness in dead cells which have been killed practically instantaneously by the action of some powerful chemical poison. Under suitable conditions it is believed that treatment of this kind fixes the constituent parts of the nucleus in very nearly the same relative positions as they occupied in life at the moment immediately preceding the death of the cell. The tissues containing the cells to be examined are then usually cut into very thin sections, and other chemicals are applied to them, the result of this treatment being to stain different parts of the nucleus of different colours and with different degrees of intensity. It is to the behaviour of the structures thus made visible that our description applies, since it is impossible to follow these changes in actually living cells except to a very imperfect extent. It may be pointed out, however, that we have every reason for believing that the differential effect produced by the processes of fixing and staining only serves to render more clearly visible real differences which actually existed during the life of the cell, and some indications of many of these differences have even been actually seen in living cells under exceptionally favourable conditions.

The nucleus, when treated in the manner described, is seen to be built up of a network of branching fibrils, the meshes of which enclose a comparatively clear and hyaline substance. The fibrils of the network are made up of a material of comparatively weak staining capacity ; embedded in this substance are numerous granules of a very intensely staining material which is known as *chromatin*. There are strong reasons for believing that the chromatin of the nucleus is of special importance from the point of view of the mechanism of heredity. This reticular structure of the nucleus is indicated in a diagrammatic fashion in Fig. 19.

Further light is thrown upon the detailed structure of the nucleus by the changes which become visible during the process of nuclear division. This process, which is known as *mitosis*, we must now proceed to describe.

In the description of mitosis which follows, the account of this process has been somewhat generalized and simplified, and Figs. 19 to 26, which illustrate the phenomena, are purely diagrammatic. It is hoped that the most important features of this complicated process may be in this way rendered comprehensible ; and although in different organisms considerable variations in the details of the process are to be met with, yet in their general features all ordinary mitoses in animals and plants are believed to conform to the essential type of our description.

The first change in the appearance of the nucleus which indicates that a division is about to take place consists in a rearrangement of the chromatin network, which now takes on the appearance of a tangled thread (Fig. 20). The outwardly-directed loops of this skein often correspond to the separate portions into which the thread eventually breaks up. The thread gradually grows shorter and thicker, and presently becomes divided into a number of pieces which are known as *chromosomes*. In the chromosomes the shortening and thickening process is continued until these bodies arrive finally at the form of stumpy rods, each of which often becomes bent into

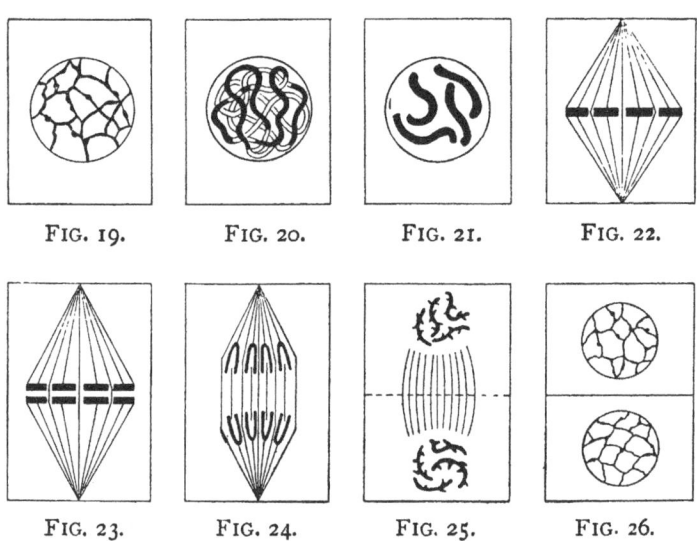

FIG. 19.　　　FIG. 20.　　　FIG. 21.　　　FIG. 22.

FIG. 23.　　　FIG. 24.　　　FIG. 25.　　　FIG. 26.

the form of a horseshoe. Meanwhile the nuclear membrane breaks down, so that the hyaline substance of the nucleus becomes continuous with that of the cell body surrounding it. A fresh phenomenon now becomes visible. A spindle-shaped arrangement makes its appearance consisting of a number of minute fibrils which connect together two points—the poles of the spindle — situated at opposite ends of the cell. The chromosomes now change their position so that they come to lie in the plane of the equator of the spindle, and about this time, but sometimes earlier, each chromosome splits longitudinally into two equal portions (Figs. 22, 23). This splitting in the case of each chromosome takes place in the equatorial plane of the spindle, so that one member of each pair of daughter chromosomes faces towards one pole of the spindle, and the second towards the other pole. The members of each pair of daughter chromosomes now begin to move away from one another towards the two poles of the spindle, and as they do so the first indication of a dividing wall between the two new cells begins to make its appearance in the equatorial plane.

Arriving at the poles, the daughter chromosomes begin to elongate, and to put out processes which finally meet and fuse with those of their neighbours to form the chromatin reticulum of the new nuclei (Fig. 25). Surrounding each new nucleus, thus developing at either pole of the now rapidly disappearing spindle, a new nuclear membrane makes its appearance ; the dividing wall in the position of the equator of the spindle develops into a complete partition (at least in the case of plants, in which, however, a number of minute passages are left penetrating the cell wall and preserving the communication between the protoplasmic contents of the separate cells) ; and the division into two new cells is thus completed (Fig. 26). Each new cell is provided with a nucleus into which has entered precisely its fair share of the chromatin which was present in the parent nucleus.

A great deal of evidence has recently accumulated to show that chromosomes are very definite and important organs. In the first place, the number of chromosomes which make their appearance at each cell division is the same in all the cells of any given creature, and this numerical constancy further extends to the cells of all the members of a particular species, though in members of allied species the number of chromosomes may be different. In widely separated species the number of chromosomes varies considerably; thus from 2 to 200 have been counted in the case of various different members of the animal and vegetable kingdoms. One of the commonest numbers found is twelve, and this number occurs in a considerable variety of different animals and plants.

Next it has been shown that the chromosomes which arise at the beginning of a nuclear division are identical with those daughter chromosomes of the preceding division which originally entered into the nucleus now about to divide. An example of the kind of evidence upon which this conclusion is based may next be given.

Figs. 27, 28, and 29 show the three possible arrangements of the four chromosomes which are found in the cells of the worm-like animal Ascaris, as seen from the direction of the pole of the spindle in the dividing nucleus. Of these arrangements, that shown in Fig. 29 is much the least common. Now in this particular case the chromosomes, when they first make their appearance immediately before the process of division, are found with their extremities situated in little pockets or bulgings of the nuclear membrane, so that their exact position is very definitely marked; and the arrangement of the chromosomes may be any one of those already indicated. Boveri observed that

FIG. 27. FIG. 28. FIG. 29.

in the case of two neighbouring cells which had originated by the division of the same mother-cell, the chromosomes made their appearance in both cases in the uncommon position of Fig. 29. Figs. 30 and 31 indicate their actual arrangement. The conclusion to be drawn from this observation is that the same

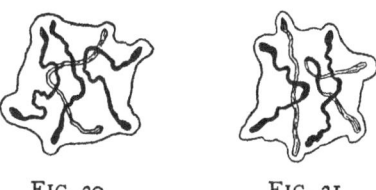

FIG. 30. FIG. 31.

chromosomes have preserved their individuality right through the resting stage of the nucleus, to reappear in the same position at the outset of a new phase of division.

It is believed, then, that the same stages which the chromosomes passed through at the close of one nuclear division, giving rise to the nuclear reticulum in the daughter nucleus, are repeated in the reverse order at the outset of the next division ; the same processes are withdrawn into the same chromosomes, and these shorten into structures identical with those which passed into the nucleus at its first formation, except that they have increased in bulk during the interval.

Boveri, in fact, concludes that the separate chromosomes are to be looked upon as distinct individuals—almost as separate simple organisms—which preserve their individuality throughout the history of the cell, and reproduce themselves, just as cells and nuclei do, by a process of bipartition. As far as the chromosomes themselves are concerned, their typical or resting form is that of the short simple rods seen in mitosis. The branched anastomozing character seen during the stage of the nuclear reticulum is associated with the active co-operation of the chromosomes in the physiological processes going forward within the nucleus. For this reason the term 'resting stage' applied to this condition of the nucleus is a particularly inappropriate one.

Boveri illustrates the amount of credence which he would attach to this theory of the individual persistence of the chromosomes throughout the resting condition of the nucleus, by means of the following analogy : ' We make water from oxygen and hydrogen, and from this water we can obtain oxygen and hydrogen again in the same proportions. Just in the same way as the chemist on the evidence of these facts regards water as containing oxygen and hydrogen, although the properties of these substances are completely in abeyance, so I believe it to be with equally good reason that our theory regards the individual chromosomes as being preserved in the resting nucleus.'*

Since Boveri expressed this opinion Rosenberg has produced further evidence of an equally convincing kind. He finds that in the case of certain plants the chromosomes do not pass over into a continuous reticulum during the resting condition of the nucleus, but remain separate, so that the same number of chromatic bodies can be counted during this stage as during the actual process of mitosis.

Boveri has also produced evidence to show that different chromosomes play different parts in the economy of the organism. For example, when different chromosomes were artificially removed from the nucleus of an embryonic cell by taking advantage of certain abnormal methods of division, the embryos which arose from these cells developed to different extents and in different abnormal ways.

This result is of particular interest, because it gives full corroboration to the suspicion, previously entertained, that the chromosomes are specially concerned with hereditary processes—with the building up of particular parts of the developing organism into shapes which resemble those of the corresponding parts displayed by other members of the same species; and it seems further to show that particular chromosomes

* Dr. T. Boveri, 'Ergebnisse über die Konstitution der Chromatischen Substanz des Zellkerns,' p. 22.

may be specially concerned in the development of particular parts.

Sutton has recently shown that the different chromosomes contained in the same nucleus of a particular animal may be of different shapes and sizes, so that each is individually recognisable. It was thus possible to demonstrate that an identically similar set of chromosomes appeared at each of several successive cell divisions. In this way additional evidence is afforded of the individual persistence of the chromosomes and of their separate identity.

We have already pointed out how, in the process of fertilization, the two conjugating germ-cells, as well as the nuclei which they contain, become completely fused together to form a single cell containing only one nucleus. It might have been expected that the separate chromosomes contained in the conjugating nuclei would also fuse together in pairs during this process, but this is not the case. The paternal and maternal chromosomes remain separate, so that the nucleus of the zygote contains twice as many chromosomes as does that of either of the gametes by the fusion of which it arose. This double number of chromosomes reappears at every cell division during the embryonic history of the zygote, and thus the fact is accounted for that the number of chromosomes in a somatic nucleus is always even.* Thus we see that the chromosomes derived from the two parents are present side by side in the nuclei of the offspring, and reproduce themselves by bipartition at every nuclear division which takes place

* See, however, p. 253 for an exceptional case.

in the zygote. In this way every somatic nucleus of the latter contains a double set of chromosomes, half of them being descended from the chromosomes introduced by one parent, whilst the other half came from the second parent.

There is reason to believe that the set of chromosomes derived from one parent is complete in itself, containing everything necessary for the development of a normal individual. Indeed, in some cases of parthenogenesis (development of the unfertilized egg), egg cells have been known to develop which contained only a single set of chromosomes. Boveri proved very prettily that the paternal set of chromosomes is equally adequate for complete development. By dint of violent shaking Boveri contrived to remove the nucleus from the egg-cells of a sea-urchin, and he afterwards allowed a sperm-nucleus to enter the enucleated egg, which presently developed into a complete embryo. Thus it was shown that the paternal as well as the maternal set of chromosomes is sufficient by itself to determine the proper production of all the organs of the embryo. But Boveri also showed that if any chromosome ot the paternal (or maternal) set were wanting in such a case, normal development of the embryo could no longer take place. Let it once more be emphasized that the somatic cells of an ordinary organism contain a double complement of essential nuclear material.

Since the gametes contain only half as many chromosomes as the somatic cells, and since the number of chromosomes present in the latter is constant for each species, it follows that either during the formation of the gametes, or at some one or other of the cell divisions leading up to their formation, there must occur a reduction in the number of chromosomes to one-half of their former number. In the case of the higher animals this reduction takes place during the two cell divisions which directly lead up to the formation of the gametes themselves. In plants, on the other hand, the reduction takes place during the formation of those cells which are known as *spores*. From these, after a certain number of intervening cell generations, the gametes themselves take their origin. These intervening cell divisions in plants are characterized in every case by the appearance of the reduced number of chromosomes. In the higher plants, in fact, a generation is, as it were, interposed between the reducing division and the actual formation of the gametes. For the spores are themselves unicellular reproductive bodies like the gametes, but differ from the latter in the fact that they develop without undergoing conjugation, and give rise to a larger or smaller mass of tissue consisting of cells with the reduced number of chromosomes. From the fact that the cells of this gamete-bearing generation contain half as many chromosomes as those of the spore-bearing generation with which it alternates, the generation produced from the spores has been spoken of as the x-generation in contrast with the ordinary, or $2x$-, generation. In animals the x-generation is reduced to a single generation of cells only, which is represented by the gametes themselves.

We must next proceed to examine the actual

method by which the reduction in the number of the chromosomes is brought about.

The simplest type of the process of reduction of the chromosomes takes place at the formation of the male germ-cells, or spermatozoa, of animals. For the sake of clearness we shall consider the case of an animal in which the somatic cells contain four chromosomes only, and in which the reduced number characteristic of the gametes is therefore two.

The reduction in number of the chromosomes takes place during two successive cell divisions which immediately lead up to the formation of the germ-cells. A particular mother-cell divides twice in rapid succession, and the four cells thus arising develop into spermatozoa without further subdivision. During these two nuclear divisions the somatic number of chromosomes becomes halved, giving rise to the number characteristic of the gametes.

Immediately before the first of these divisions the chromosomes become closely associated together in pairs, and in certain cases it has been shown that one member of each pair is very probably the descendant of a chromosome derived from the male parent, whilst the other member of the pair is the descendant of the corresponding maternal chromosome.

This association of the chromosomes in pairs may be so close, and may take place so early, that when these bodies are first visibly differentiated only half the usual number of them is to be seen. But in these cases, too, it is reasonable to believe that each of the chromosomes actually visible consists of a maternal

16

and a paternal member fused together. Each of the visible chromatic bodies next divides into four parts, the set of four deeply staining bodies being known as a *tetrad*. Thus when there are four somatic chromosomes the number of tetrads appearing will be two (Fig. 32). A mitosis now takes place, during which there is no further division of chromosomes, but half of each

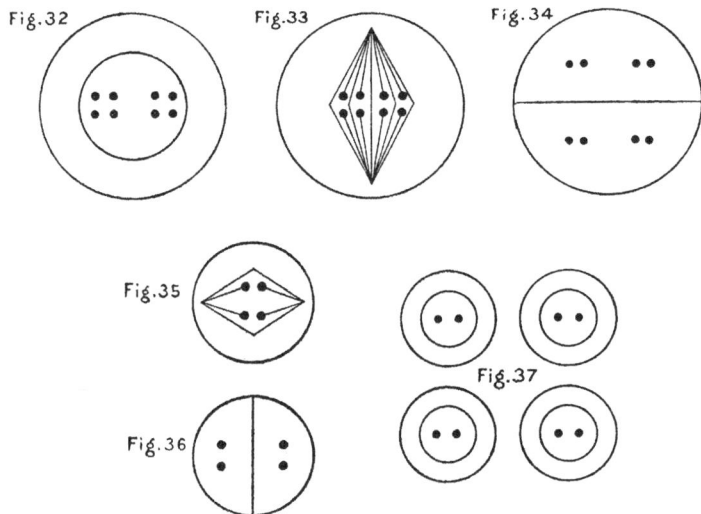

tetrad passes to either pole of the nuclear spindle, so that each daughter nucleus comes to contain two half-tetrads, each consisting of a pair of deeply-staining bodies (Fig. 34). This division is not followed by the production of a resting nucleus, for before any nuclear reticulum is formed, and while the half-tetrads still retain their definite appearance, the daughter nuclei divide again. At this second division in each nucleus the separate members of each of the two half-tetrads pass to opposite poles (Figs. 35, 36).   In the nucleus of each of the four cells which thus arise there is, therefore, present one quarter of each of the four chromosomes which originally appeared—one member, that is to say, of each tetrad (Fig. 37).   Each of the cells of which we have thus traced the origin develops directly into a single spermatozoon.

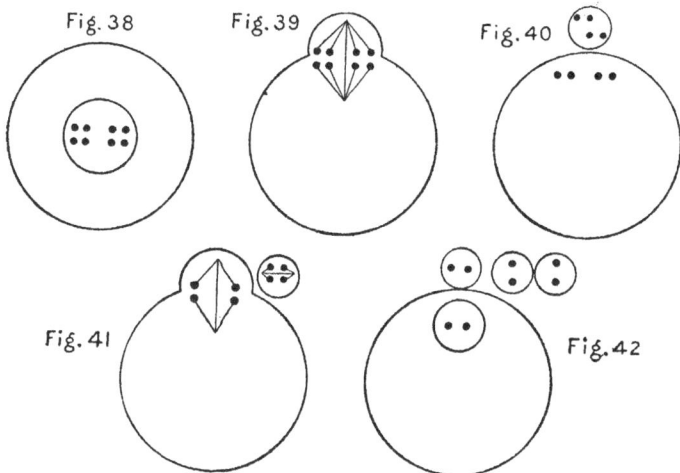

Fig. 38   Fig. 39   Fig. 40   Fig. 41   Fig. 42

The method of development, or maturation, of the ova, or egg-cells, of animals is in all essential respects similar to the process by which the spermatozoa arise. It differs, however, in the fact that of the four cells which result from the corresponding divisions, one is very large and constitutes the ovum, whilst the other three are very minute, and are apparently of no further importance.   In the accompanying diagrams (Figs. 40 to 42), the smaller cells, or polar bodies, have been

16—2

enormously exaggerated relatively to the size of the ovum itself.

The original tetrad is believed in all cases, and has been actually observed in a few cases, to arise by a separation of the two fused chromosomes, followed by a division of each of these bodies into two.   In cases where the chromosomes retain their rod-like appearance throughout these changes there would seem to be some doubt as to whether the first of the divisions giving rise to the ' tetrad ' is transverse or longitudinal in direction, and it is possible that the process may be different in different cases.   But it is generally agreed that the first division separates the two original chromosomes, and that at the first of the two nuclear divisions which ensue the members of a pair of parental chromosomes pass into separate nuclei.   The second division, on the other hand, like an ordinary mitosis, separates halves of chromosomes.   This agreement among authorities is explained by the circumstance that those observers who have seen a longitudinal first division believe that the parental chromosomes conjugated side by side, whilst those who describe a transverse division describe also an end-to-end conjugation of the chromosomes.

The first of these two ideas is the one illustrated in the accompanying diagram (Fig. 43), representing the behaviour of a single pair of parental chromosomes during the two nuclear divisions which give rise to four sperm cells.   The chromosome derived from one parent is shaded, whilst the other is left blank.

Thus the first of the two gamete-producing divisions

differs from all other mitoses in the fact that in it an
actual separation of whole chromosomes takes place ;
it is a qualitative and not only a quantitative division.
It is to this mitosis that the term *reducing division* is
properly applied.

We have to notice that at one stage of the process
now described the chromosomes derived from the two
parents are in a close state of fusion. It would seem
as if the actual conjugation of chromosomes, which
failed to take place when the conjugating gametes and
their nuclei fused together in the formation of the
zygote, was only delayed, and now occurs hundreds or

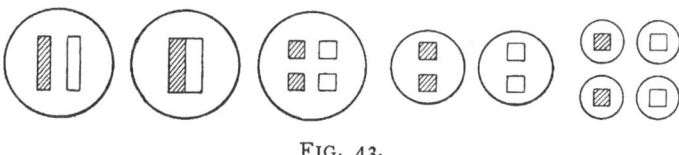

FIG. 43.

thousands of cell generations after the actual process
of fertilization, and immediately before the production
of those cells which are to give rise to the new
individual.

It may be pointed out that, although the chromosomes which emerge from this fusion seem to be
identical with those which entered into it, yet it is
difficult to believe that they have not undergone some
change, or exercised some mutual influence upon one
another. If no such influence has been exerted, it is
difficult to imagine any possible reason for the process
of fusion taking place at all.

In the higher plants a similar reducing division

takes place at the formation of the spores, which arise in sets of four, each set corresponding to a group of four spermatozoa, or to the ovum and the three polar bodies of an animal. In the case of flowering plants the nuclei contained in the spores make a few further divisions, at each of which the reduced number of chromosomes is to be observed, and one or more of the cells thus finally produced take on the character of germ-cells. The spores are of two kinds, large and small, the latter being the pollen grains. The larger spores give rise to female gametes and the smaller to male, and fertilization takes place in the ordinary manner by a fusion between the nuclei of these germ-cells.

We have seen so far that the number of chromosomes contained in the somatic nuclei of a given species is always the same, and is always even. We have also seen that this number is made up of two separate sets derived respectively from the two parents, and that the members of the two sets preserve their separate individuality right through the long series of nuclear divisions which take place during the development of the individual zygote. A fusion of chromosomes of paternal and maternal origin respectively takes place only in the direct line of ancestry of the germ-cells which are destined to give rise to new members of the species. This process of fusion takes place in animals immediately before the formation of the actual germ-cells, but in plants a larger or smaller number of cell generations earlier. After fusion the paternal and maternal chromosomes apparently separate, and the nuclear division which ensues differs from all other mitoses in the fact that instead of merely dissevering halves of chromosomes, the actual somatic chromosomes separate and become distributed equally between the resulting nuclei; so that in these nuclei, and in the germ nuclei which arise by their division, the number of chromosomes is reduced to half the somatic number. When fertilization takes place the somatic number of chromosomes is restored by the union of nuclei, each of which contains half that number.

Is it possible to throw any further light upon the meaning of these facts regarding the behaviour of the minute constituent parts of organisms?

Let us return to Mendel's experimental discovery, of which an account was given in the last two chapters, and let us consider the case of a cross between parents which differ in respect of two pairs of allelomorphs. Expressing these pairs as A - a and B - b, Mendel showed that the germ-cells of the cross-bred or heterozygote bear in equal numbers the combinations AB, Ab, aB, and ab. Now, it seems clear from this behaviour that the allelomorphs must be represented in the cells of the organism by some kind of definite particles, which remain distinct from one another throughout all the cell divisions of the body, since we know that at the formation of the germ-cells these characters are capable of becoming completely segregated. Let us, then, trace the behaviour of the allelomorphs in a diagrammatic way, regarding each as a distinct particle. These particles we may distinguish by certain letters. A and a are the allelomorphs of one pair,

B and *b* those of the other, and we will suppose that one of the parents exhibits the characters *A* and *b* and the other the characters *a* and *B* (Fig. 44). Then, in the zygote resulting from fertilization, *A*, *a*, *B*, and *b* will all be present.

Since all the cells, at least in the direct line of ancestry of the gametes, must contain every allelomorph, it will be necessary for the particle representing each allelomorph always to divide into two before a cell division takes place, for only in this way can something corresponding to each allelomorph pass into each of the two cells produced by the division. And a similar process will be repeated at each somatic mitosis (Fig. 44). At the formation of the germ-cells, however, or at some preceding cell division, the two members of each pair of allelomorphs must become separated from one another in such a way that the particles originally derived from different parents pass over into different cells. When two pairs of allelomorphs are concerned, this process of separation can take place in either of the two ways shown in Fig. 45. And the experimental evidence shows that the two methods occur with equal frequency in the formation of the germ-cells of the same heterozygote.

Anyone who has succeeded in following the above account of the behaviour of the supposed particles representing Mendelian allelomorphs in the cells of a hybrid organism, on comparing it with the preceding description of the behaviour of chromosomes in the somatic and reducing divisions respectively, can scarcely fail to be struck by the extraordinary simi-

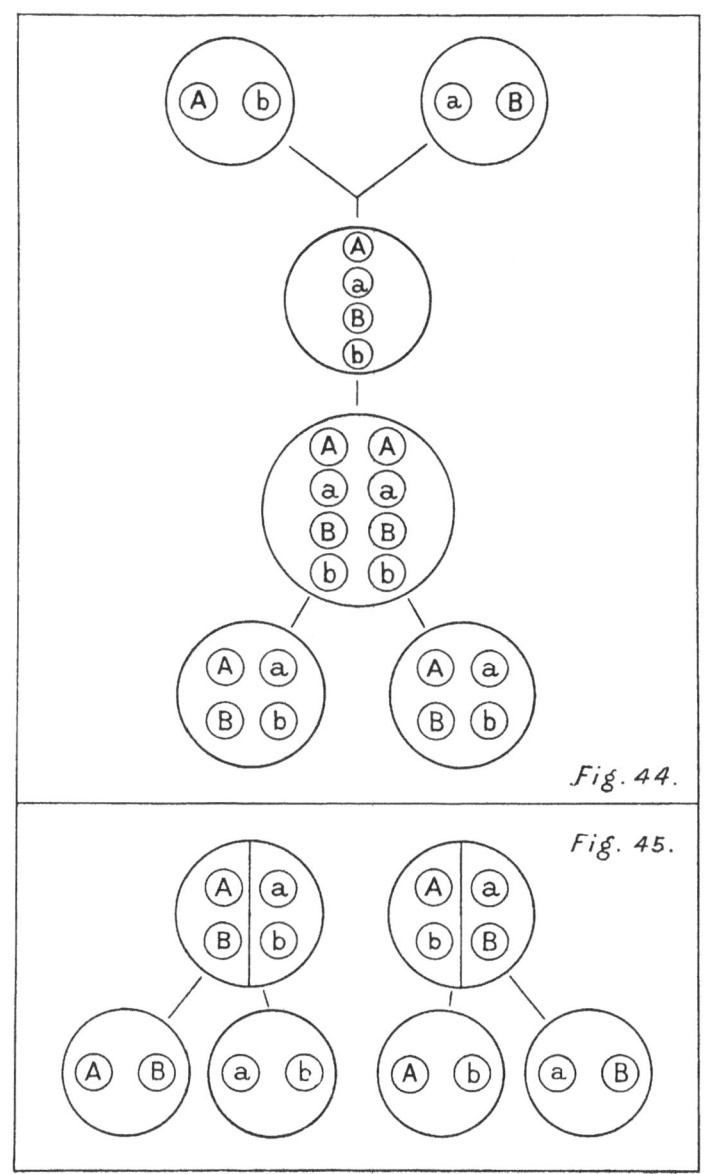

Fig. 44.

Fig. 45.

larity between the two processes. It seems quite clear that there must be some real connection between the behaviour of chromosomes as seen microscopically on the one hand, and the behaviour of allelomorphic characters as deduced from the results of experiment on the other ; and that the evidence derived from these two forms of study is bound to be of considerable mutual benefit.

At first sight it might be thought that the chromosomes are the actual bearers of Mendelian characters, in the sense that each chromosome represents a single allelomorph ; and, indeed, there is no fundamental difference between the behaviour of chromosomes and that of our supposed character-bearing particles. But there is, at least in some cases, a fatal objection to this belief in the fact that in certain plants the number of separate allelomorphic pairs which may be born by a hybrid is greater than the reduced number of chromosomes which the germ-cells of this hybrid contain. For instance, in the case of the pea the reduced number of chromosomes is seven, and Mendel himself described the behaviour of seven independent pairs of allelomorphs in peas. Recent study has revealed the presence of at least four additional pairs of allelomorphs in these plants, all of which are probably equally independent of one another.

We must, therefore, seek a different explanation, and de Vries has recently suggested one which up to the present time appears the most likely to represent the true account of the phenomena. De Vries' explanation is associated with the finer structure of the chromosomes themselves, a subject upon which we have not hitherto entered. Under high powers of the microscope, and after very careful preparation, it is possible to observe that each chromosome contains a number of separate darkly-staining granules which are known as *chromomeres*. When the pairs of parental chromosomes fuse together previous to the reducing division, the chromomeres which they contain appear to meet together in corresponding pairs. The members of each pair fuse together completely, afterwards separating as the chromosomes separate.

FIG. 46.

De Vries supposes the Mendelian allelomorphs to be contained in the chromomeres, and that when these granules fuse together an exchange of allelomorphs takes place between the chromosomes. This exchange proceeds in such a way that when the chromosomes separate after fusion, it is a matter of simple chance whether a particular allelomorph has remained in the chromomere which originally contained it, or has passed over into the other member of the pair. Thus, in a sufficient number of cases we should get all possible chance distributions of allelomorphs between the two chromosomes, except that, of course, the two members of the same pair of allelomorphs would never coexist in the same chromosome. Since the two chromosomes of a pair pass into different germ-cells, precisely that chance distribution of allelomorphs which is required on the Mendelian theory would thus be arrived at.

De Vries' explanation throws light on one phenomenon which is not accounted for on the supposition that each chromosome represents a separate allelomorph. In the diagrams previously given of the behaviour of Mendelian characters within the cells we have given no indication of a conjugation in pairs previous to the reducing division. Such a process of fusion is, however, one of the most marked phenomena in the behaviour of the chromosomes at the parallel stage of their existence. On the chromosome-allelomorph view, the phenomenon of mitosis as bringing about an equal division of hereditary particles between the cells, and the process of reduction in the number of the chromosomes, are both accounted for, but there is no explanation of the fusion between the pairs of chromosomes. On de Vries' view, however, this process is necessary in order to bring about the necessary redistribution of allelomorphs between the chromosomes, and so between the germ-cells into which the latter pass.

In cases where the phenomenon of correlation or coupling has been observed we must suppose that there is some mechanism which causes the representative particles of the respective characters concerned to remain in company during the process by which the other allelomorphs are being reassorted between the chromosomes. Of this process of coupling the cytologists have not yet been able to observe any visible indication in the behaviour of the chromosomes, any more than they can really see the redistribution of the supposed factors carried by the chromomeres. But apart from this it must be allowed that the facts of experiment and of microscopic observation fit in with one another in a remarkable way, and that the Mendelian theory throws considerable light on the minute features of cell anatomy.

The possibility still remains that in certain cases particular characters may be associated with particular chromosomes as a whole, and we shall next proceed to describe what actually seems to be an example of this sort.

The case we have to describe is directly concerned with one of the most interesting and elusive of biological problems—namely, the problem of the heredity of sex. Prof. E. B. Wilson has recently investigated the behaviour of the chromosomes in the somatic cells and in the germ-cells of a particular species of insect known as *Protenor belfragi*. The case afforded by this animal is remarkable, inasmuch as the somatic cells in the male, and only in the male, contain an odd number of chromosomes. An irregularity is accordingly introduced into the process of fusion of the chromosomes in pairs, which, as already described, always precedes the formation of the germ-cells with their reduced number of chromosomes. In the case of the male *Protenor* all the chromosomes fuse in pairs except one, which is, of necessity, left over. This odd chromosome is described as the *heterotropic* chromosome. The female *Protenor* has one more chromosome in its somatic cells than the male, thus making up an even number—that is to say, in the female the pair to the odd chromosome of the male is present, so that there are two heterotropic chromosomes, or *idiochromosomes*. These fuse and separate in the reducing division, which thus proceeds in the normal manner in this sex. In the male, on the other hand, when the reducing division occurs, the heterotropic chromosome passes complete into one of the resulting cells. In the second gamete-producing division, every chromosome present having divided into two, the products of this division pass into different gametes. These latter divisions are of two kinds, since in one of them the heterotropic chromosome takes part, whilst in the other it is wanting ; consequently, two out of the four spermatozoa eventually produced contain the heterotropic chromosome and two do not. (Only one of each kind is shown in Fig. 47.) Thus there is a differentiation of the spermatozoa into two different kinds, and one of these kinds contains a chromosome less than the other. On the other hand, all the eggs (as well as the polar bodies) contain an idiochromosome.

In fertilization some of the eggs become impregnated by spermatozoa containing the heterotropic chromosome. Such eggs invariably develop into females having a pair of idiochromosomes in each somatic cell. Other eggs are fertilized by spermatozoa lacking the heterotropic chromosome, and these become males, their somatic cells containing only the single heterotropic chromosome derived from the egg. The ac-

companying diagram illustrates the behaviour of the
chromosomes during these processes. The heterotropic chromosomes are represented as black, whilst
the remaining chromosomes are left white, and for the
sake of simplicity only two pairs of the latter are
indicated in the somatic cells.

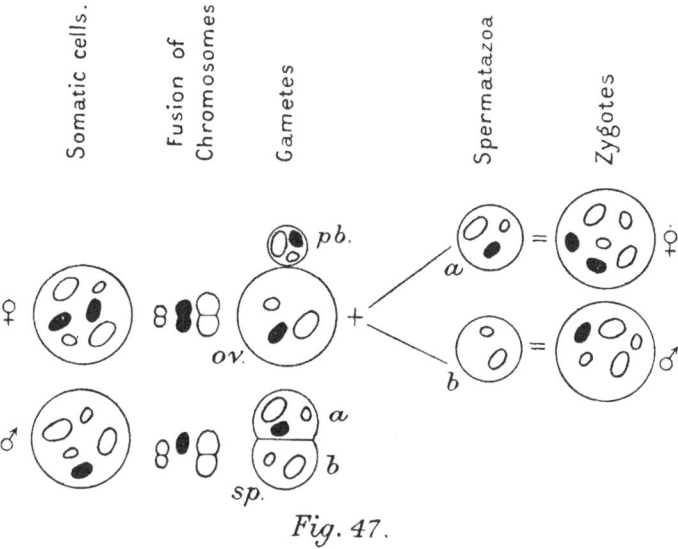

Fig. 47.

ov, ovum ; *pb*, polar body ; *sp*, spermatozoa (*a* and *b* the two kinds).

When the above facts are taken into consideration
it is scarcely possible to doubt that there is a causal
relationship between the characteristics of the female
sex and the presence of two heterotropic chromosomes, and that a similar connection exists between
maleness and the presence of only one. Let us trace
this relationship a little further.

The facts clearly prove, in the first place, that the unpaired heterotropic chromosome alternates between the two sexes in alternate generations, passing from the male to the female in the production of females, and from the female to the male in the production of males (see the diagram).

Assuming that these particular chromosomes are really concerned in the determination of sex, Wilson suggests the following interpretation on Mendelian lines. Since the heterotropic chromosome is the only one present in the male, it must represent the male determinant. But, since spermatozoa which contain this chromosome produce only females, the maternal mate of the male heterotropic chromosome, already present in the egg, must be a dominant female determinant. And in the process of fertilization which gives rise to males the heterotropic chromosome derived from the egg must represent the male determinant. Two different sorts of eggs are therefore produced—presumably in equal numbers—which contain the male and female determinant respectively ; the former are fertilized only by spermatozoa lacking the heterotropic chromosome and *vice versâ*. The combinations which arise in this way may be represented as (*m*)*f* and *m*.

A selective process of fertilization is therefore a *sine quâ non* for this explanation—it must be impossible for a spermatozoon bearing the male determinant to fertilize an egg in which a male determinant is already present—in other words, only eggs containing the female determinant can be fertilized by sperms which contain a heterotropic chromosome. At first sight this necessary supposition of a selective fertilization presented itself to Wilson as a serious difficulty. He points out, however, that the experiments of Cuénot with yellow mice* afford perfectly independent evidence of the actual occurrence of a selective fertilization of this kind in a particular species of animal (and similarly Wilson's observations lend a welcome confirmation to Cuénot's conclusions from his experiments).

In mice Cuénot found that on crossing together heterozygous yellows $CYCG \times CYCG$, he obtained no pure dominant yellows—$CYCY$—such as were to be expected ; and this in spite of the fact that $CYCG \times CGCG$ gave equal numbers of yellows and greys (showing that the $CYCG$ individuals were giving off equal numbers of yellow- and of grey-bearing gametes). Cuénot explained his unusual result by supposing that yellow-bearing spermatozoa unite only with grey-bearing eggs, and *vice versâ*. On this explanation, one might at first sight expect to get a proportion of two yellows to one grey instead of nearly three to one, which was the ratio actually observed, since the fertilization yellow by yellow fails altogether. But Wilson points out that, since spermatozoa are in great numerical excess as compared with eggs, it will be possible for all the Y-bearing eggs to be fertilized by G-bearing spermatozoa, as well as half the G-bearing eggs by Y-bearing spermatozoa, thus bringing the proportion of yellows to greys more or less nearly up to three to one.

* See p. 195.

In another species of insect closely allied to *Protenor* the somatic cells of the male, like those of the female, contain each a pair of idiochromosomes; but in the male one member of the pair is much larger than the other, whilst in the female they are of equal size. The behaviour of the larger member of the unequal pair of chromosomes, in the various nuclear processes which occur during the life-history, is precisely like that of the single heterotropic chromosome of *Protenor*. It is still possible to regard this chromosome as representing a recessive male determinant, and to suppose that the process of sex determination is precisely similar in the two cases. On this supposition, the smaller idiochromosome is regarded as being without function so far as sex is concerned.

In a third insect belonging to the same natural group both male and female sexes bear alike a pair of idiochromosomes of equal size. Here, again, it is possible to apply the same theory of sex determination by simply disregarding one of the idiochromosomes of the male as unimportant. We may suppose, in fact, that one of these chromosomes corresponds to the smaller idiochromosome of the preceding case, and that it takes no essential part in these phenomena. The fact that this chromosome takes no active part in these processes may, indeed, have led to its reduction in the second of the three species, and to its final disappearance in the first.

Thus, by dint of a good deal of speculation, Wilson has arrived at a possible Mendelian description of the phenomenon of sex in a species in which the chromosomes of male and female are alike ; and it is a description which has its basis in actual phenomena observed in two other related animals.

Other explanations of the sex phenomena of *Protenor* and its allies may, as Wilson points out, be pos· sible ; but if Cuénot's result meets with confirmation from other observers, the one given above will certainly take the rank of a very probable hypothesis. In any case these observations of Wilson's mark a considerable advance along the road towards a complete interpretation of the problem of sex determination. One thing, at any rate, seems certain, and that is that the male or female character is already fully determined in the fertilized egg, so that no subsequent action of the environment can have any influence upon the sex of the offspring. This definite determination of sex in the very earliest stage of the zygote follows of necessity from any Mendelian view of the phenomenon, and the evidence afforded by *Protenor* points very clearly in the same direction.

By way of further illustrating the far-reaching importance of the information which has been rendered available by the combined use of experimental and cytological methods, we may here briefly criticise the celebrated theory of inheritance put forward by Weismann in 1892 under the name of the ' Germ Plasm Theory.' Some notice of this theory, which might otherwise have been permitted to go the way of similar valuable provisional hypotheses, is rendered almost necessary by the circumstance of its having been recently revived in a prominent manner in the English translation of Weismann's book, 'The Evolution Theory.' In this book, published in 1904, the bearing of the Mendelian evidence upon the subject of inheritance is practically ignored; although, in the face of the definite experimental information now rendered available, the younger biologists, at least, are beginning to realize that the circumstantial evidence, formerly so much relied upon, will in future constitute a much less prominent feature in these discussions.

Weismann's theory of inheritance, and the Theory of Ancestral Heredity in its original form, are based upon a common assumption, which is now shown by Mendel's discovery to have been unfounded. This is the assumption that all ancestors of the same degree— *e.g.*, grandparents—make a substantially equal contribution to the hereditary qualities of the offspring. Mendel has shown that in the case of particular hereditary characteristics this is not the case.

But if we venture to criticise Weismann's conception in the light of more recent knowledge, it must not be forgotten that biology, and especially modern cytology, owes a great debt to Weismann. To Weismann is due the conception of the isolation of the germ-cells from somatic influences, a view which is in complete accordance with the Mendelian view of the inheritance of definite characters. And it was Weismann who first emphasized the belief that the chromosomes represent those parts of the nucleus which are specially concerned in the processes of heredity. These conceptions— which, indeed, constitute an essential part of his own theory of heredity—have stood the test of time in an admirable manner.

Let us turn our attention, then, for a short space to the Germ Plasm Theory of inheritance. On Weismann's theory, as in most other theories of heredity from the time of Darwin and Nägeli downwards, the separate parts of the living organism are supposed to be represented by separate material particles in the germ-cells. These representative particles are known as *determinants*. A complete set of determinants in which every part of the organism is thus represented constitutes an *id*. So far Weismann's hypothesis is in close agreement with the idea of representative particles which we are driven to adopt by the facts of Mendelian inheritance, except that, following de Vries, we should speak of separate characters rather than parts as being thus represented ; for there seems to be no doubt that the same character-determinant can affect the development of a number of different parts. But at the next step the Mendelian parts company with Weismann. The latter assumes that the cells of an organism contain a large number of *ids*, or complete sets of determinants, half of the total number being derived from either parent, and that, although at the reducing division which precedes the formation of the gametes the total number of *ids* is reduced to half of what it was in the somatic cells, still, several *ids* derived from each parent are present in every germ-cell.

Thus the reduced number of chromosomes in the germ-cells is regarded as containing all the primary constituents of both parents. And it is an essential point in Weismann's theory that he regards a given germ-cell as containing a considerable number of *ids* derived from its ancestors, all near ancestors being thus represented.

But Mendel's experiments and others of the same kind show, in the case of a great number of different characters, that although every essential character is represented in every germ-cell, yet each Mendelian character is represented by a paternal or a maternal determinant only, and not by both. Thus, not only are all immediate ancestors not represented in the germ-cells in respect of any particular character, but only one of the parents is so represented—to the complete exclusion, so far as we can tell, of the other parent. In fact, we are led to believe that the germ-cells contain one set of determinants only—a single *id*—whilst the somatic cells contain two *ids* only. The Mendelian theory is thus seen to be considerably simpler than the germ plasm theory, which it replaces. At the same time it must not be forgotten that many of the conceptions used in the Mendelian expression of the facts are borrowed from Weismann's theory, and that but for Weismann's work it would have been impossible for us to have got so far in the co-ordination of the facts derived from experiment and microscopic observation respectively.

The preceding sketch may serve to show how Mendel's observations have been found to throw light upon many of the facts of cytology the meaning of which was previously obscure ; and how it affords at the same time a criterion by which may be tested the truth of theories based upon the interpretation of minute phenomena only made visible by the highest powers of the microscope. The disinterment of Mendel's discovery took place only six years ago ; and the rapid manner in which the facts of cytology have been found to fall into line with Mendelian conceptions augurs well for the future progress of discovery in these fields.

CHAPTER X

IN the preceding chapters a considerable variety of topics has been dealt with, and in spite of the fact that all are more or less intimately connected with the study of organic evolution, the nearly historical order in which the subject-matter has been in great part presented has inevitably rendered the treatment a little disjointed.

The method we have so far adopted serves to illustrate the state of transition in which our studies stand, and which it is our first object to assist in hastening to a close—the transition between the speculative philosophy of evolution and the exact science of genetics.

Future treatises on genetics will make a fair beginning with the law of Mendel, and will then deal with the application of this law in detail ; and in this concluding chapter we may adopt the same method, and proceed to show how Mendel's discovery affords the connecting-link between the various divergent branches which we have already sketched in outline.

The central generalization, then, around which the subjects considered in the preceding pages are found naturally to group themselves is afforded by the law of inheritance discovered by the Abbé Mendel about the year 1865. This discovery has rendered possible that rapid advance of the science of genetics, or the study of the hereditary phenomena of organisms, which has taken place during the first few years of the twentieth century. There can be no sort of doubt that Mendel's brief paper is the most important contribution of its size which has ever been made to biological science. Little apology is therefore needed for formulating once again the law based by Professor Correns upon the conclusions which this paper contains.

Mendel's law relates to the inheritance of certain definite characters, which have since been called allelomorphs. It is a distinctive feature of allelomorphic characters that they are found to group themselves naturally into pairs of more or less antagonistic qualities. In many cases the pair is represented by the presence and absence respectively of a certain definite feature. The two allelomorphs of a pair may be conveniently written as A and a.

We have seen that the cells of zygotic organisms—organisms, that is to say, which have arisen by the process of sexual reproduction—contain a double complement of hereditary qualities. Such cells may contain A and A, a and a, or A and a. The forms AA and aa are described as homozygotes, the form Aa as a heterozygote. In the simpler cases we are enabled to study the behaviour of such a single pair of allelomorphs by itself, without reference to any other features which the animals or plants under consideration may display. The demonstration that there exist definite and separable unit characters of this kind is the first great debt that science owes to Mendel.

Up to the present our certain knowledge of the Mendelian behaviour of unit characters has been confined to cases of cross-breeding. In the simplest case which we have to consider, two homozygote forms, AA and aa, are crossed together.

The external character or visible appearance of the heterozygote Aa, produced in this manner, differs in different cases. In the commonest case A represents the dominant allelomorph, and in this case the appearance of the heterozygote Aa is practically indistinguishable from that of the homozygote AA. In other cases the heterozygote Aa is different in appearance from either homozygote AA or aa. Sometimes Aa is intermediate between AA and aa, in other cases it is to all appearances totally distinct from either.

So much for the external appearance of homozygote and heterozygote forms. In the production of the gametes, or germ-cells, we arrive once more at the simplest possible form of hereditary constitution, for we believe each feature in the body to be represented in the germ-cells by a single determining factor only. Still confining our attention to the representatives of a single pair of allelomorphs, we find that the germ-cells of a homozygote contain only A or only a, as the case may be. But in the case of the germ-cells derived from a heterozygote, A and a are represented in an equal number of the gametes produced by the same individual. And the separation between the two allelomorphs is found in almost all cases to be perfectly complete.

This complete segregation of the two allelomorphs in equal numbers of the germ-cells of a heterozygote constitutes the first and most important section of the generalization known as Mendel's law.

The second part of the law refers to the fact that, as a general rule, separate pairs of allelomorphs segregate quite independently of one another. To this rule a few exceptions have been recorded in cases where apparently distinct pairs of determining factors behave in segregation like a single pair of allelomorphs. In such cases we regard the members of the distinct pairs of allelomorphs as being coupled together, although no serious attempt has yet been made to picture the way in which this coupling comes about. In other cases, again, the coupling seems to be only partial. These phenomena are not yet by any means completely understood.

The fact that in the great majority of cases separate pairs of allelomorphs segregate independently of one another leads to the possibility of new combinations of the parental characters being formed in the germ-cells of the cross-bred individuals ; in fact, this must always happen when the parent types differ in more than one pair of segregable characters. When two similar germ-cells, each bearing the same new combination of allelomorphs, meet together in fertilization, the result is a new zygotic combination which is a pure type in respect of the characters concerned, and henceforth breeds true. Thus if *AB . AB* is crossed with *ab . ab* the heterozygote *AB . ab* produces in equal numbers the germ-cells *AB, Ab, aB*, and *ab*. Among the combinations of these germ-cells which are represented by the various offspring of the heterozygote there must appear *Ab . Ab* and *aB . aB*—novel types which are pure in constitution, and which may form the starting-points for new strains or races.

Upon this fact depends the enormous importance of Mendel's law in the breeding of new and useful types of animals and plants.   When it is remembered that in wheat, for example, resistance and non-resistance to the attacks of disease, earliness and lateness of ripening, good and bad milling quality, are all pairs of Mendelian allelomorphs, and that it is now possible to take a different example of these qualities from each of three different strains, and to combine them together in a single new variety with perfect certainty and in four generations, it does not require much imagination to foresee that every department of the animal and plant breeding industries must sooner or later benefit enormously from Mendel's discovery.

So far we have only been dealing with the very simplest of Mendelian phenomena, leading to the arithmetical addition and subtraction of definite visible characters. Other kinds of allelomorphs also exist which undergo a similar process of segregation during gamete formation, following Mendel's law in a perfect manner ; but which may remain entirely invisible and unsuspected so long as certain other allelomorphs, belonging to quite distinct pairs, are excluded from the zygotes in which these invisible factors are concealed.   When this other complementary allelomorph is introduced, however, by crossing with an individual which contains it, the feature previously hidden becomes visible, giving rise to the phenomenon which has long been familiar under the name of reversion on crossing. The demonstration of these invisible factors, and of the fact that they also obey Mendel's law with perfect regularity, is surely one of the most remarkable discoveries which have ever been made in the whole history of biology. This, again, is a piece of knowledge which may be of the very greatest importance, not only to breeders of bright flowers, some of which are already known to exhibit the phenomenon described, but also in all classes of breeding work where similar facts doubtless await discovery.

To the man of science, however, the practical aspect of these achievements will be of little account in comparison with the importance of their application to the advance of human knowledge in that most fascinating of scientific studies—biology. Let us, then, turn to consider the way in which Mendel's discovery affects other branches of biological science.

We have so recently had occasion to point to the remarkable coalition between Mendelism and cytology that little more need be said here upon the subject. Mendel's theory has, indeed, thrown a flood of light upon the meaning of the microscopic phenomena exhibited by the minute constituent parts of the cells of living organisms, phenomena the meaning of which could only be vaguely guessed at previously to the introduction of the new method.

The intimate connection between Mendelism and cytology rests to a large extent upon the close parallel which exists between the behaviour of allelomorphic characters on the one hand and that of chromosomes on the other.

In the germ-cells of the higher animals the allelomorphs of the Mendelian become segregated, being reunited in fertilization, and, as a consequence, the cells of the zygote contain twice as many of these factors as do the gametes or germ-cells themselves.

Similarly, in the cell processes upon which the vital functions of the higher animals are founded, the number of chromosomes characteristic of somatic or zygotic cells becomes halved at the formation of the gametes, the double number being restored by the association of chromosomes derived from two separate gametes in the process of fertilization. We have said that in the higher animals the gametes are sometimes spoken of as constituting an ' x-' generation, which alternates with the ' $2x$-' generation represented by the zygote. We may justify the use of these expressions by a brief comparative statement of the facts relating to the two so-called generations which recur in the life-history of certain families of plants. In doing so we shall begin our account with the most primitive and simplest forms, and then pass on to other types which are regarded as standing on higher planes of evolution.

What are probably some of the most primitive members of the vegetable kingdom belong to the class of the green algæ. This group includes a great number of comparatively lowly organisms, the majority of which dwell submerged beneath the surface of fresh or salt water. In such members of the green algæ as have so far been examined from this point of view, it would appear that the $2x$-generation is exclusively represented by the single cell which arises as the actual product of conjugation between a pair of gametes. Reduction takes place in the actual zygotic cell, so that each of the products of this cell's division shows once more the reduced number of chromosomes. Thus the great bulk—the vegetative mass—of the species is constituted by the x-generation, and the $2x$-generation is composed of a single cell only—a state of things which is exactly the reverse of what is to be seen in the higher animals.

In the vegetable kingdom evolution seems to have been accompanied by a gradual increase of the $2x$-generation, and a corresponding reduction of the x-generation in point of importance. Between the two extremes afforded by the algæ on the one hand, and the flowering plants on the other, we can trace a series of intermediate stages represented by types in which many other features also must be regarded as standing on intermediate planes of organization.

As an example of an intermediate condition of this kind, we may take the case of the ferns.

The fern plant, as commonly understood, represents the $2x$-generation. The method by which the life-history of the fern plant is continued is by the formation of unicellular reproductive bodies which are known as spores. The formation of the spores takes place in sets of four, and their production is preceded by a reducing division, so that each spore nucleus contains half as many chromosomes as the nuclei of the fern-plant—the spores, in fact, represent the initiation of the x generation.

Spores take no part in any process of conjugation. They at once germinate and enter on an embryonic development of their own, giving rise to a considerable mass of cells, all of which contain the reduced number of chromosomes. Thus in the case of the fern we have a small but well-developed x-generation alternating with a much larger $2x$-generation. The mass of cellular tissue making up the x-generation has been named the *prothallus*.

Certain cells of the prothallus develop, without change in the number of their chromosomes, into the gametes. These are differentiated in the usual way into male and female—ova and spermatozoids respectively.

Fertilization of the ovum by the spermatozoid gives rise to a zygote in which the double number of chromosomes is restored. In this way the $2x$-generation or fern plant is initiated, and by the usual processes of cell multiplication and differentiation this body becomes completed, developing its characteristic fronds and so forth. Thus in the ferns the $2x$-generation has arrived at a high degree of development, and represents the chief bulk of the plant. The x-generation, however, still embodies a considerable mass of cells.

Turning to the higher plants, among which we may include those which produce typical flowers with stamens or with pistils, or more usually with both, we find that the x-generation has become still further reduced, so that it no longer occupies an independent phase of the life-history, but has come to be entirely dependent upon the $2x$-generation for its support.

A plant which bears both stamens and pistils gives rise to spores of two kinds, differing greatly in size. The smaller spores are represented by the pollen-grains, and in these, after one or two cell divisions, unaccompanied by growth, the one or two male gametes are produced. The small association of cells arising in this way is all that is left of the x-generation on the male side.

The nucleus of the larger spore also divides a few times, and one of the final products of division becomes the ovum. Spore and ovum, as well as the few intervening cells, bear the reduced number of chromosomes. The x-generation thus represented is never set free, but remains enclosed in the tissues of the $2x$-generation right up to the time of fertilization. In the process of fertilization the double number of chromosomes characteristic of the $2x$-generation is once more arrived at.

We can look upon the $2x$-generation of the higher plants as being formed by an expansion of the fertilized ovum. The zygote, instead of comprising a single cell only, by dint of delaying the reducing division, has come to consist of a great mass of cells, all the nuclei of which contain the double number of chromosomes. This fact is also our excuse for applying the same term of zygote to the cell produced by the conjugation of

18

gametes, as well as to the mass of cells to which the zygote (in the strictest sense) eventually gives rise. In the simplest forms, such as the algæ, the cell- and nuclear-fusion constituting conjugation are immediately followed by fusion of the chromosomes, an event which we have seen to be the first step towards a reduction in the number of these bodies. In the higher plants, by delaying this fusion of chromosomes until many cell generations later than the fusion of the nuclei, the advantages associated with the possession of a double nucleus have been obtained for a large and complicated mass of cells. And this mass has gradually advanced in organization and relative importance, until ultimately the x-generation has been reduced almost to the vanishing point.

The sex-phenomena of the higher animals can most readily be brought into line with those of the higher plants if we consider that in animals the spore and the gamete are identical ; the x-generation is here condensed into the smallest possible limits—namely, those of a single cell.

A female animal produces ova, and a male produces spermatozoa. Similarly, we may regard as a female plant one which produces only the larger variety of spores from which ova arise ; and we may regard as a male plant one which produces only pollen. It is much more usual to find a flowering plant bearing both pistils and stamens, and producing both large and small spores. Such an organism is described as hermaphrodite—bearing both sexes. Among animals examples of hermaphrodite species are also not infrequent, and here, just as in the case of plants, whole families may display this method of reproduction.

We see, then, that the course of evolution in the vegetable kingdom would appear to have been accompanied by a gradual increase in the $2x$-generation at the expense of the x-generation. Starting with lowly marine organisms, and passing upwards through the mosses and ferns to the flowering plants, we find a steady diminution in the x-generation, whilst the vegetative labour of the plant is taken over by the $2x$-generation. It is, therefore, proper to suppose that organisms in which the main stage in the life-history is of double origin, and bears a double complement of hereditary factors, have some advantage over organisms in which this is not the case. We cannot, of course, be certain as to the exact nature of this advantage, but we may point out that it is only in the former kind of organisms that the operation of Mendel's law can lead to the production of new combinations of parental characters in the body which represents the main stage of the life-history ; and that this circumstance may possibly lead to a greater power of adaptability to external circumstances.

Perhaps the most interesting application of the information afforded by Mendel's discovery is shown in its bearing upon the question of discontinuity in the origin of species. The fact of the definite and discontinuous inheritance of the differentiating features which distinguish cultivated varieties from one another would point very plainly to a belief that such differences had arisen in a definite and discontinuous manner, even if we did not actually know from direct evidence that the origin of new races under cultivation is usually sudden and complete.

It is not necessary to repeat Darwin's demonstration of the close analogy between the origin of varieties under cultivation and the origin of species in Nature. It is more to the purpose to point out that Mendel's law has already been shown to hold good in the case of many differences which have certainly not arisen under cultivation, and that we have, moreover, sure knowledge of the definite and spontaneous origin of some natural species.

Here we arrive at a point at which the evidence is not yet by any means complete. We do not know whether all or even many specific differences obey Mendel's law on crossing, and a sharp limit is put to our researches in this direction by the fact that so many natural hybrids are sterile. Still less do we know from direct evidence whether the majority of natural species have arisen discontinuously, although there is much circumstantial evidence which points to the conclusion that this must have been the case.

Clearly this discontinuous method of variation is likely to repay some further discussion. That such mutation, or definite variation, is a phenomenon of the germ-cells follows from the fact that every germ-cell normally bears the complete specific character. Bateson has shown that we must regard mutation as consisting in the production of new kinds of gametes, which differ from those normally characteristic of the species. Such a change is most readily pictured by imagining an asymmetrical nuclear division taking place immediately before the formation of the germ-cells, and this would lead us to expect a mutating species to give rise to more than one new kind of offspring at the same time. Such was actually the case with the *Œnothera Lamarckiana* studied by de Vries; and this observation stands as the most complete piece of evidence of a mutating species so far known to us. We may be assured, then, that the complete potential nature of new types as well as of old ones is already laid down in the germ-cells previous to fertilization. As Bateson puts it : ' For the first time in the history of evolutionary thought Mendel's discovery enables us to form some picture of the process which results in genetic variation. It is simply the segregation of a new kind of gamete, bearing one or more characters distinct from those of the type. We can answer one of the oldest questions in philosophy. In terms of the ancient riddle, we may reply that the owl's egg existed before the owl ; or, if we hesitate about the owl, we may be sure about the bantam.'*

Let us consider a little more closely the evidence of mutation afforded by de Vries' studies of *Œnothera Lamarckiana*. Semi-wild specimens of this species, when transplanted and carefully observed, were found to yield nearly 3 per cent. of seedlings which differed definitely from their parent, and among these *mutants* some fifteen distinct new sorts were described. Some of the new species equalled or even surpassed the parent

* British Association, Cambridge, 1904. Address to the Zoological Section, p. 14.

O. Lamarckiana in vigour and prolific habit, and two of them actually became established side by side with the parent type without man's assistance.

It is unfortunate from the point of view of de Vries' interpretation of this case that the behaviour of *O. Lamarckiana* should suggest in some respects, as Bateson has pointed out, the phenomena of hybridization. It must be observed in support of de Vries' view that the species appears to exhibit the same phenomenon in other localities, and, further, that it has not been possible to make any suggestion as to the second species with which the pure *Lamarckiana* might be supposed to have been crossed.

From one point of view, as de Vries has himself pointed out, mutation in *Œnothera* is clearly a phenomenon of hybrids, and this circumstance of itself introduces considerable complications into the story.

We saw just now that there is every reason for the conviction that mutation takes place in the germ-cells, and not in the zygote after fertilization. Since the number of mutants given off under the most favourable circumstances did not exceed 3 per cent. of the total offspring, the enormous majority of mutated germ-cells (on de Vries' view) must unite with germ-cells bearing the ordinary specific character. Consequently, the new types which appear will in most cases have originated in the form of a cross between a mutated germ-cell and an ordinary germ-cell. And since this is not the final limit to the possible complications of the case, we can easily recognise that the complete interpretation of the behaviour of *Œnothera Lamarckiana* is not by any means an easy matter.

As enunciated by de Vries, the theory of mutation amounts to a very complete and definite hypothesis. A large part of this author's suggestions are, however, almost purely speculative, and for this reason we have treated the whole at somewhat less length than it perhaps deserves. Some of de Vries' speculations are, indeed, more picturesque than convincing.

Thus, de Vries regards the number of unit characters —each of which has arisen by a single mutation—to be quite limited, even in the highest organisms. Three or four thousand such characters, he thinks, may go to build up the hereditary endowment of the most complicated species. He further supposes a period of mutation to recur about once in 4,000 years. Four thousand multiplied by 4,000 gives 16,000,000—the number of years required to evolve the lords of creation from a ' primordial protoplasmic atomic globule.' And he points out that this estimate is well within the limits of geological time as allowed by the physicist. In this way de Vries believes that his mutation theory removes a difficulty which besets the selection hypothesis—the difficulty, namely, of insufficient time. The selectionist may reasonably reply that the amount of change necessary to produce in 4,000 years, by the gradual method, a difference equal to that represented by a single unit character, might very well be quite imperceptible in a single generation.

We may summarize our present conclusions as to

the discontinuous nature of species in the following manner : A great number of specific characters are, without doubt, definite ; they are inherited as definite entities, and there can be no question that their first coming into existence was a definite event. Every year tends to increase the range of characters to which the conception of discontinuity has to be applied.

Certain groups of characters do, however, seem to exhibit the phenomena of continuity. Let us endeavour to arrive at some closer idea as to the nature of these characters.

A study of continuous variations very quickly leads to the conclusion that the variable features are those which are especially liable to modification during the lifetime of the individual, owing to the action of external circumstances. Such quantitative features of size and shape and number of parts are particularly plastic in the case of plants.

The habit, or general form and appearance, of a plant is a feature very characteristic of individual species. The presence of a dwarf or of a tall habit does, indeed, constitute a frequent distinction between different strains of garden plants, and the inheritance of these characters in many cases follows Mendel's law. But leaving aside this particular example, the inheritance of habit is very little understood ; although habit is a feature which is very liable to considerable fluctuations. Habit seems, in fact, usually to afford an example of continuous variability.

The habit of some species of plants when grown under alpine conditions on mountain summits is so different from that of the same species when growing in the plains, that inexperienced persons might readily suppose two such forms to belong to as many distinct species. At intermediate levels the habit is more or less intermediate. Bonniér made the experiment of dividing individual plants into two portions, planting one part at a high elevation and the other near the level of the sea. In a few years the plant grown on the mountain had taken on the full alpine habit, whilst that grown on the plain retained the ordinary appearance of the species. In this way very considerable differences in habit were shown to be directly dependent on external conditions.

In some few cases the environment determines the production of perfectly definite and discontinuous features. The water ranunculus, when growing submerged beneath the surface of a pond, produces leaves the blades of which are cut up into a great number of fine thread-like segments. As soon as the top of the plant reaches the surface of the water those leaf rudiments which are just commencing their existence proceed to develop in a totally different fashion. The leaves to which they give rise possess a wide and undivided blade, which floats upon the surface of the water. The two sorts of leaves are as utterly different in appearance as it is possible for leaves to be. Yet the effect of external conditions upon the young leaf-rudiment determines which of the two kinds is to appear.

In this instance we see a discontinuous change in conditions—the change from water to air as a surrounding medium—giving rise to a discontinuous change in structure. Such cases are, however, comparatively rare. Much more usually the changes in external conditions are continuous, as changes of altitude, moisture, or chemical composition of the soil, and so on ; and the changes induced by them in the plant are similarly of a continuous kind.

In most animals changes in external circumstances have a much smaller influence on the form and structure of the individual than is the case with plants. In animals considerable modifications are, however, brought about by exercise and the use of different parts, as Lamarck long ago observed. But these modifying factors usually affect all the members of a single species in nearly the same manner. Nevertheless, some part of the differences between individuals in respect of strength and of proportion, and possibly also of stature, is undoubtedly associated with differences of training and nutrition, as the example of the human race is sufficient to show. Professor Cope has pointed out how the proper development of such structures as the joints of vertebrates depends to a very large extent upon exercise ; and the effect of disuse may be practically tested by anyone whom accident obliges to keep a knee or other joint immovable for any length of time. The so-called play in which the young of many animals indulge—for example, lambs and kittens —must have a great influence upon the perfection of their locomotory functions.

We can now see more clearly the reason for that great instability of vegetative type which sessile animals, like plants, exhibit. No necessity for definite and co-ordinated movements involving their whole structure forces the development of these animals along certain definite paths. External circumstance is, therefore, free to mould them into a host of slightly different shapes. And thus the great variability of the species of corals, for instance, is doubtless determined to a large extent by the influence of different environmental conditions.

Strictly speaking, the term variability ought not to be applied to modifications of this description. It will, perhaps, be most convenient, however, to distinguish true variations—having their origin in differences among the germ-cells—as *genetic variations*, contrasting them with the *acquired variations* which arise during the development of individuals.

Enough has now been said to show that it is a very difficult matter to distinguish in the case of continuous variations between those which are genetic and those which are acquired.

It is easy to understand how acquired variations come to be continuous, and to obey the law of normal variability. We saw that the normal distribution of characters was induced by the random operation of a multitude of small causes. During the development of the individual a great number of different external influences come into play, leading to slight modifications of every part, now in one direction, now in another. This being so, we may be quite sure that a large proportion of the normal variability which any species exhibits is acquired.

Now we saw that there seems to be good evidence that normal or continuous variations are inherited. Logic does not, however, permit us to make the step: Acquired variations are continuous variations; continuous variations are inherited; therefore acquired variations are inherited. It seems, indeed, to be this fallacy which has led to the long-continued belief in the inheritance of acquired characters as an important factor in organic evolution in spite of so many arguments to the contrary.

Formal disproof of this proposition is very difficult, and in the meantime the confusion between continuous acquired variations and continuous genetic variations, which is always present in practice, constitutes a very serious drawback to the biometric method of research. At present Johannsen's explanation of these phenomena seems to afford so much the simplest solution that we may once more repeat his statement of the case, though with the proviso that the proof of his hypothesis is still to be awaited.

Johannsen looks upon a population which, as a whole, exhibits continuous or normal variability, as being capable of analysis into a number of pure lines. In a single pure line genetic variability is sensibly absent. The members of such a pure line exhibit, however, very considerable acquired variability, so that in this way each line shows a normal variability of its own. And the range of this variability may greatly exceed the limits which separate two pure lines from one another. The result is to give a completely blurred picture when all the lines are looked at simultaneously. And thus the normal variability of the population as a whole is brought about by the combination of these two separate factors.

This statement applies to the case of an organism in which self-fertilization is the general rule, so that in this way the separate lines are kept distinct. Where cross-fertilization takes place between the members of different pure lines the case becomes enormously complicated, and this is much the most frequent instance which we have actually to deal with. It has been suggested that the members of different lines when crossed together may display Mendelian phenomena, but the existence of so large a proportion of acquired variability renders the problem of analyzing the result almost insuperable. We have seen, however, that the numerical results obtained by the biometricians do not appear to be inconsistent with the existence of Mendelian inheritance in populations.

We find, then, that the questions of inheritance of acquired characters and of evolution by the aid of continuous genetic variations are not yet absolutely settled. But the evidence is certainly such that for all practical purposes the former factor at least may be disregarded. Meanwhile the number of cases in which discontinuity of inheritance can be shown to hold good is constantly increasing, and the analysis of some cases of supposed continuous variation into discontinuous Mendelian factors has already been made. It may be safely concluded that a very large part, if not the whole, of evolution has taken place by the discontinuous method.

New little species—Jordan's species—arise, then, from time to time, each at a single step, from pre-existing species. Upon the material thus supplied natural selection operates ; the weaker go to the wall, the stronger survive. This is also, in all probability, the way in which adaptations have arisen. Creatures which came into existence displaying a particular new structure, which happened to be fitted for a particular new function or suited to a particular nitch in Nature, survived and flourished exceedingly. Those in which undesirable organs appeared perished and were no more seen. To take Aristotle's example. If a man were to be born with molars in front and incisors at the back of his jaw he would die—at least, in the days before dentistry. Having his teeth in the positions in which they actually stand (although not for this reason only), he survives and rules the world.

After all, the difference between the point of view thus briefly indicated, and that of Darwin as expressed in the ' Origin of Species,' is only one of detail—of detail as to the particular sort of variations by which evolution chiefly proceeds. Darwin's analogy between the origin of species in Nature and the origin of races under cultivation may be repeated with emphasis, although Huxley's famous criticism, to the effect that races which are sterile together have not arisen in cultivation, is not yet completely answered. But this renders the discontinuous origin of such sterility only the more likely ; and when we recall the Mendelian behaviour of such characters as long and short style in the primrose, or sterility of the anthers in the sweet-pea, the solution of the problem does not seem very far to seek.

Let us see how the principles of which an outline has now been given affect the human race itself. The question of improving the human stock in this country has lately excited a good deal of attention. But without a scientific knowledge of the factors upon which improvement and degeneration depend the discussion is not likely to be of much profit, and in such a case misdirected energy may be even worse than apathy. Without venturing to make any very positive suggestions, it may at least be pointed out that our present practice in these matters is in almost every case the very worst possible.

Professor Karl Pearson has lately shown how the low birth-rate of the professional and middle classes— the classes amongst which the intelligence of the nation is to a large extent segregated—leads to the recruiting of these classes from amongst the lower and less intelligent strata of society. In other words, a steady breeding out of intelligence is taking place. Recognising that intelligence is an important factor in national greatness, we proceed to remedy this defect by endeavouring to reduce the infant mortality among the less desirable classes, and by offering every inducement to the production of large families by the said lower strata of society ; indeed, we propose to remove from them all responsibility for the production of children, and to feed and house the latter as we already educate them (save the mark !) at the expense of the State.

The principles of heredity teach us that education and training, however beneficial they may be to individuals, have no material effect upon the stock itself. If they have any effect at all, this is undoubtedly unimportant in comparison with the effect which would be produced by the selection of individuals which exhibit desirable qualities. The demand for a higher birth-rate ought to apply strictly to desirables. Instead of this the cry is for education and physical training, processes which can have no permanent beneficial effect upon the race.

One writer who holds to some extent the attention of the intelligent public has recognised the true state of affairs—I mean Mr. Bernard Shaw. Unfortunately the public does not take Mr. Bernard Shaw seriously, wherein, when I recall Mr. Shaw's published views on such topics as vivisection and the medical profession, the said public has my sympathy. Nevertheless I know of no better expression of the moral to be drawn from the science of genetics than that which is embodied in the following passage :

' I do not know whether you have any illusions left on the subject of education, progress, and so forth. I have none. Any pamphleteer can show the way to better things, but when there is no will there is no way. My nurse was fond of remarking that you cannot make a silk purse out of a sow's ear, and the more I see of the efforts of our churches and universities and literary sages to raise the mass above its own level, the more convinced I am that my nurse was right. Progress can do nothing but make the most of us all as we are, and that most would clearly not be enough even if those who are already raised out of the lowest abysses would allow the others a chance. The bubble of heredity has been pricked, the certainty that acquirements are negligible as elements in practical heredity has demolished the hopes of the educationists as well as the terrors of the degeneracy-mongers, and we now know that there is no hereditary " governing class " any more than a hereditary hooliganism. We must either breed political capacity or be ruined by democracy, which was forced on us by the failure of the older alternatives. Yet if despotism failed only for want of a capable benevolent despot, what chance has democracy, which requires a whole population of capable voters—that is, of political critics who, if they cannot govern in person for lack of spare energy or specific talent for administration, can at least recognise and appreciate capacity and benevolence in others, and so govern through capably benevolent representatives ? Where are such voters to be found to-day ? Nowhere. Promiscuous breeding has produced a weakness of character that is too timid to face the full stringency of a thoroughly competitive struggle for existence, and too lazy and petty to organize the commonwealth co-operatively. Beingc owards, we defeat natural selection under cover of philanthropy ; being sluggards, we neglect artificial selection under cover of delicacy and morality.'*

Mr. Shaw recognises, however, that our knowledge is at present insufficient to prescribe for the breeding

* ' Man and Superman,' p. xxiii.

of a ' Superman,' even if we were able to come to any agreement as to what qualities are the most desirable. Nevertheless it is along the lines which we have endeavoured to indicate that such knowledge must be sought in the future.

GLOSSARY

[*Many technical terms not included in this glossary are printed in italics on their first appearance in the body of the book, and their meaning is then defined. Such definitions may be discovered on a reference to the index.*]

ADAPTATION. — A teleological explanation of the correspondence often shown between the structure and habits of a particular creature and the environment in which the creature lives.

ALBINO.—An animal or plant characterized by the absence of colouring matter from its external tissues.

ALGÆ.—A group of plants, mostly aquatic and of relatively simple organization.

ANTHER.—The upper part of a stamen, containing the pollen.

ATOM.—The smallest part of a chemical element which can exist as such.

AXIL.—The angle enclosed between the base or stalk of a leaf and the stem upon which the leaf is borne.

BINOMIAL NOMENCLATURE.—The application of a double name to an animal or plant, the first name being that of the genus, the second that of the species.

BIOLOGY.—The science of the phenomena of life.

BIOMETRY. — The application of statistical methods to biological problems.

BOTANY.—The scientific study of plants.

CALYX.—The outermost whorl of floral leaves, which in the bud usually encloses the other organs of the flower.

CHARACTER.—In heredity, a single definable attribute.

CLASS.—One of the larger subdivisions of the animal kingdom—*e.g.,* mammals, birds.

COMPOSITÆ.—A family of plants, including the daisy, chrysanthemum, and many others.

CONJUGATION.—The process of fusion of a pair of gametes.

COROLLA.—The second envelope of a flower, consisting of petals—leaf-like organs—usually brightly coloured.

CORPUSCLE.—A very minute particle.

CYTOLOGY.—The scientific study of the minute constituent parts of organisms by the aid of the microscope.

DENUDATION.—The wearing away of the earth's surface by the action of rain, rivers, etc.

DIFFERENTIATION.—The separation or discrimination of parts which were previously more or less united and uniform.

EMBRYO.—A young plant or animal—usually one which is still contained in the seed or the womb.

EMBRYOLOGY.—The history of the development of young plants or animals from the egg.

ENVIRONMENT.—Natural surroundings.

EVOLUTION.—See p. 21.

FAMILY.—A group of allied genera, as the family of apes (*Anthropoidæ*), the buttercup family (*Ranunculaceæ*).

FAUNA.—The sum total of animals inhabiting a particular region.

FERTILIZATION.—The union of male and female reproductive cells or gametes.

FLORETS.—The separate flowers of a crowded inflorescence.

GAMETES.—Sexual cells which unite in conjugation or fertilization.

GENUS.—A group of allied species.

GEOLOGY.—The study of the earth's crust.

GEOMETRIC RATE OF INCREASE.—Progress consisting in successive multiplications of the preceding number, instead of simply in additions to it.

GERM-CELLS.—See GAMETES.

HERBALIST.—One who collects and studies herbs.

HEREDITY.—The transference of similar characters from one generation of organisms to another, a process effected by means of the germ-cells or gametes.

IGNEOUS.—Produced in connection with great heat.

INBREEDING.—The mating together of near relatives for a number of generations.

LARVA.—The young of an insect after it has emerged from the egg—*e.g.*, a caterpillar.

MANTISES.—A group of predatory insects.

MAXILLARY.—Connected with the mouth parts.

MORPHOLOGY.—The study of form and structure.

MUTATION.—The sudden origin of a new species at a single step.

ORGANISM.—A living creature.

ORNITHOLOGIST.—A student of birds.

OVARY.—In animals the organ which produces ova. In plants the organ which contains the ovules.

OVUM.—The female gamete.

OVULE.—The structure surrounding the spore which gives rise to the female gamete or ovum in the higher plants.

PETAL.—One of the (usually) coloured leaves composing the corolla.

PETALOID.—Resembling the corolla, usually in the circumstance of being coloured.

PHYSIOLOGY.—The study of the functions of organisms.

PIN-EYED.—Having the stigma on a level with the throat of the corolla, and the anthers lower down, enclosed within the tube.

PISTIL.—The central organ of a flower, which contains the ovules, and ultimately becomes the fruit, or the chief part of it.

POLLEN.—Those spores of the flowering plants which produce the male gametes.

POLLINATION.—The transference of pollen to the stigma of a plant.

PRIMARY, SECONDARY, AND TERTIARY EPOCHS.—The three great divisions of geological time during which the known fossiliferous strata were deposited.

RADICAL LEAVES.—Leaves arising immediately from the root-stock in the form of a rosette.

REVERSION.—The reappearance in the offspring of a character proper to a more or less remote ancestor, and not exhibited by the immediate parents.

ROTIFERS.—A kind of minute aquatic animals.

SEGMENT.—One of a series of more or less similar transverse divisions.

SESSILE.—Fixed and stationary, but (in the strict sense) without a stalk.

SOMATIC.—Belonging to the body of a zygote.

SPECIES, LINNÆAN. — A group of organisms of closely similar appearance.

SPECIES, JORDAN'S.—A group of organisms believed to have arisen by a mutation. (Jordan himself did not, however, suppose so.)

SPORT.—A marked mutation—often one occurring under domestication.

STAMENS.—The organs of a flower which bear the pollen.

STANDARD.—The large, upright petal at the back of such a flower as that of the sweet-pea.

STIGMA.—The uppermost part of the pistil, upon which the pollen is received.

STRATUM.—A layer.

STYLE.—A stalk connecting the stigma with the ovary—part of the pistil.

TESTA.—The skin or coat covering a seed.

THRUM-EYED.—Having the anthers situated at the throat of the corolla, and the stigma lower down, enclosed in the tube.

TUBE.—The basal tubular portion of a corolla in which the separate petals are closely fused together, as is the case with that of the primrose.

UNICELLULAR.—Consisting of a single cell.

VARIATION, CONTINUOUS.—See Chapter IV.
VARIATION, DISCONTINUOUS.—See Chapter V.

WINGS.—The lateral petals of a pea-flower.

ZOOLOGY.—The scientific study of animals.

ZYGOTE.—The organism produced by the fusion of a pair of gametes.

INDEX

BILLING AND SONS, LTD., PRINTERS, GUILDFORD